Progress Notes for Higher Physics

by

W. Kennedy

ISBN 0 7169 3254 7
© W. Kennedy, 2001

ROBERT GIBSON · Publisher
17 Fitzroy Place, Glasgow, G3 7SF

PREFACE

Progressive Notes fo Higher Grade Physics covers the topics studied for the SQA Higher Grade Physics syllabus. It contains **a planned sequence of notes** constructed to guide the student systematically towards specific problems.

These notes are designed on a facing page format with diagrams, graphs, worked examples, etc., to give a highly visual impact.

The notes can be used as a stand-alone resource to assist student learning. However, it is recommended that they are used in conjunction with the sister book *Progressive Problems for Higher Physics*.

♦ Every chapter in the 'H' notebook **links** with every chapter in the 'H' Problem book (e.g., Chapter 6 is Gas Laws in both books).

♦ Within the chapter each concept **links** with appropriate problems, e.g., Boyle's law notes **link** to specific problems on Boyle's law.

♦ The principal concept is one of seamless progression using progression links (see opposite).

The material has been used extensively. It has been closely monitored as to its value and relevance to the learning process. Consequently, it has been developed and refined to its present level.

W. Kennedy, 2001.

ACKNOWLEDGEMENTS

I would like to thank the physics department, staff and pupils of St. Mungo's High School, Falkirk, who trialled the materials and helped to refine them.

THE PROGRESSION LINK SYSTEM

- The notes are designed on a facing page format.

- Important relationships or equations are highlighted in the "LEARN" box.

- A **progression link** directs the student to specific problems in *Progressive Problems for Higher Physics*.

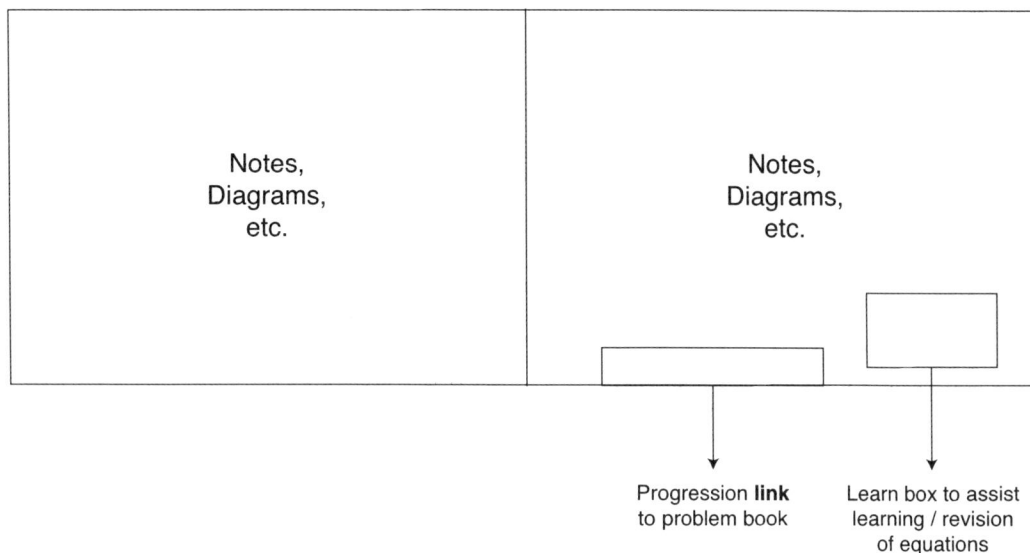

Notes,
Diagrams,
etc.

Notes,
Diagrams,
etc.

Progression **link**
to problem book

Learn box to assist
learning / revision
of equations

Students should read the notes, learn the relationship or equation in the "learn" box and progress to the problems as directed.

The teacher can select appropriate past paper questions after the pupil has gained confidence by completing the progressive problems.

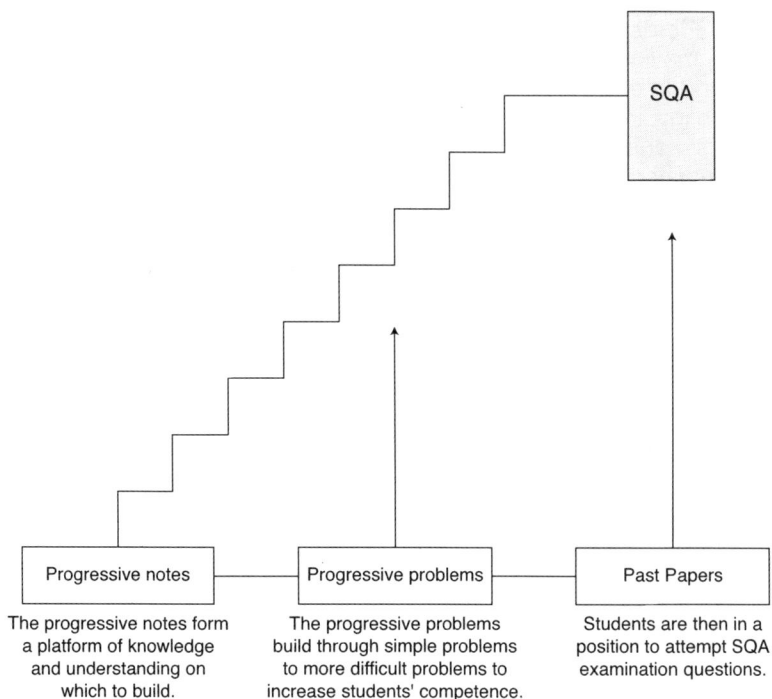

SQA

Progressive notes	Progressive problems	Past Papers
The progressive notes form a platform of knowledge and understanding on which to build.	The progressive problems build through simple problems to more difficult problems to increase students' competence.	Students are then in a position to attempt SQA examination questions.

CONTENTS

DATA SHEET

COMMON PHYSICAL QUANTITIES

Quantity	Symbol	Value	Quantity	Symbol	Value
Speed of light in vacuum	c	3.00×10^8 m s^{-1}	Mass of electron	m_e	9.11×10^{-31} kg
Charge on electron	e	-1.60×10^{-19} C	Mass of neutron	m_n	1.675×10^{-27} kg
Gravitational acceleration	g	9.8 m s^{-2}	Mass of proton	m_p	1.673×10^{-27} kg
Planck's constant	h	6.63×10^{-34} J s			

REFRACTIVE INDICES

The refractive indices refer to sodium light of wavelength 589 nm and to substances at a temperature of 273 K.

Substance	Refractive index	Substance	Refractive index
Diamond	2·42	Water	1·33
Crown Glass	1·50	Air	1·00

SPECTRAL LINES

Element	Wavelength / nm	Colour	Element	Wavelength / nm	Colour
Hydrogen	656	Red	Cadmium	644	Red
	486	Blue-green		509	Green
	434	Blue-violet		480	Blue
	410	Violet	Lasers		
	397	Ultraviolet	Element	Wavelength / nm	Colour
	389	Ultraviolet	Carbon dioxide	9550 } 10590 }	Infrared
Sodium	589	Yellow	Helium-neon	633	Red

PROPERTIES OF SELECTED MATERIALS

Substance	Density / kg m^{-3}	Melting Point / K	Boiling Point / K
Aluminium	2.70×10^3	933	2623
Copper	8.96×10^3	1357	2853
Ice	9.20×10^2	273	. . . .
Sea Water	1.02×10^3	264	377
Water	1.00×10^3	273	373
Air	1·29	. . . .	. . . .
Hydrogen	9.0×10^{-2}	14	20

The gas densities refer to a temperature of 273 K and a pressure of 1.01×10^5 Pa.

Acknowledgement is hereby given to the SQA to reproduce this data sheet.

FORMULAE LIST

UNIT 1

Average speed $= \dfrac{\text{total distance travelled}}{\text{total time taken}}$

Average velocity $= \dfrac{\text{displacement}}{\text{total time taken}}$

$a = \dfrac{v - u}{t}$

average speed $= \dfrac{u + v}{2}$

$v = u + at$

$s = ut + \dfrac{1}{2}at^2$

$v^2 = u^2 + 2as$

$W = mg$

$F = ma$

$W.D. = Fd$

$E_p = mgh$

$E_k = \dfrac{1}{2}mv^2$

$P = \dfrac{E}{t}$

$p = mv$

$Ft = mv - mu$

Density $= \dfrac{m}{V}$

Pressure $= \dfrac{\text{Force}}{\text{Area}}$

$p = h\rho g$

$p \propto \dfrac{1}{V}$

$V \propto T_K$

$p \propto T_K$

$T_K = T_{°C} + 273$

$\dfrac{p_1 V_1}{T_1} = \dfrac{p_2 V_2}{T_2}$

UNIT 2

$Q = It$

$W = QV$

$R_T = R_1 + R_2 + R_3$ (series)

$\dfrac{1}{R_T} = \dfrac{1}{R_1} + \dfrac{1}{R_2} + \dfrac{1}{R_3}$ (parallel)

$V = IR$

$V_1 = \dfrac{R_1}{R_T} \times V_{\text{supply}}$ (voltage division)

$P = VI = I^2 R = \dfrac{V^2}{R}$

$E = Ir + V_{\text{t.p.d.}}$

$\dfrac{R_1}{R_2} = \dfrac{R_3}{R_4}$

$Q = CV$

$E = \dfrac{1}{2}QV = \dfrac{1}{2}CV^2 = \dfrac{1}{2}\dfrac{Q^2}{C}$

$V_{\text{peak}} = \sqrt{2} \times V_{\text{rms}}$

$V_0 = -\dfrac{R_f}{R_1} V_1$

$V_0 = (V_2 - V_1)\dfrac{R_f}{R_1}$

Voltage gain $= \dfrac{\text{output voltage}}{\text{input voltage}}$

Power gain $= \dfrac{\text{output power}}{\text{input power}}$

UNIT 3

$v = f\lambda$

$f = \dfrac{1}{T}$

$S_2 P - S_1 P = n\lambda$

$S_2 B - S_1 B = \left(n + \dfrac{1}{2}\right)\lambda$

$d \sin \theta = n\lambda$

$n = \dfrac{\sin \theta_1}{\sin \theta_2} = \dfrac{v_1}{v_2} = \dfrac{\lambda_1}{\lambda_2}$

$n = \dfrac{1}{\sin \theta_C}$

Intensity $= \dfrac{\text{Power}}{\text{Area}}$

$I \propto \dfrac{1}{d^2}$

$E = hf \qquad I = Nhf$

$W_2 - W_1 = hf$

$hf = hf_0 + \dfrac{1}{2}mv^2$

$A = \dfrac{N}{t}$

$D = \dfrac{E}{m}$

$H = DQ$

$\overset{\bullet}{H} = \dfrac{H}{t}$

$E = mc^2$

UNIT 1

MECHANICS AND PROPERTIES OF MATTER

CHAPTER 1

VECTORS

VECTORS

Definition

A **vector** quantity is one which can only be fully described by both **magnitude** and **direction** (and the appropriate units).

A **scalar** quantity is one which can be fully described by the **magnitude** alone (and the appropriate units).

The difference can be explained by comparing distance with displacement. Displacement is the distance "as the crow flies" between the start and finish of a journey.

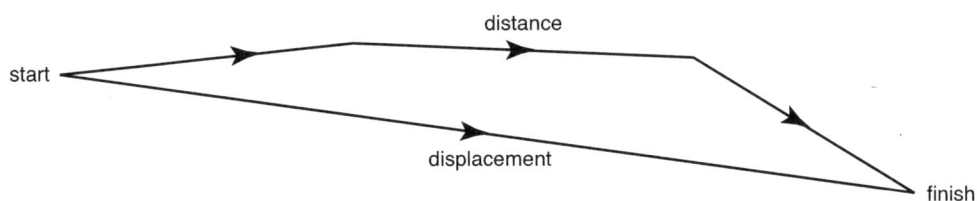

Some examples of vectors and scalars:

Scalar	Vector
distance	displacement
speed	velocity
time	acceleration
mass	force

Vectors can either be positive or negative but scalars are always positive.

Distance / Displacement and Speed / Velocity

A man walks 4 km north (in one hour) and then 3 km east (in one hour).

The two distances can be added in the following vector diagram.

The student has a choice between a scale diagram and Pythagoras:

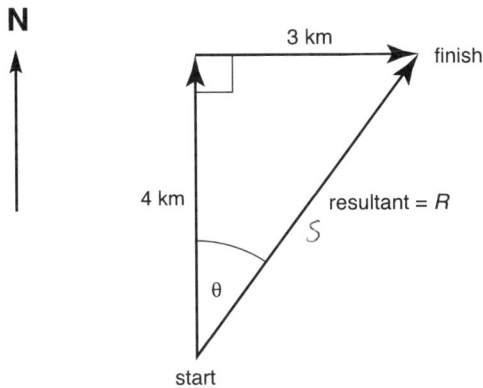

$$R^2 = 4^2 + 3^2$$

$$R^2 = 16 + 9$$

$$R = \sqrt{25}$$

$$\underline{\underline{R = 5 \text{ km}}}$$

$$\tan \theta = \frac{3}{4}$$

$$\theta = \tan^{-1} 0.75$$

$$\underline{\underline{\theta = 36.9°}}$$

Total distance gone	=	7 km
displacement	=	5 km 36.9° to the east of north [037°]

Average speed $= \dfrac{\text{total distance gone}}{\text{total time taken}} = \dfrac{7 \text{ km}}{2 \text{ hours}} = \underline{\underline{3.5 \text{ km h}^{-1}}}$

Average velocity $= \dfrac{\text{displacement}}{\text{total time taken}} = \dfrac{5 \text{ km bearing } 037°}{2 \text{ hours}} = \underline{\underline{2.5 \text{ km h}^{-1} \text{ bearing } 037°}}$

LEARN
Average speed $= \dfrac{\text{total distance travelled}}{\text{total time taken}}$
Average velocity $= \dfrac{\text{displacement}}{\text{total time taken}}$

Progress to P.P. 'H' P. Page 7, nos 1–12

Addition of Vectors

EXAMPLE 1

Problem

Two forces of 200 N pull a 5 kg mass (see diagram below).

Find:

(a)   the resultant force
(b)   the acceleration of the mass (assuming no friction).

Solution

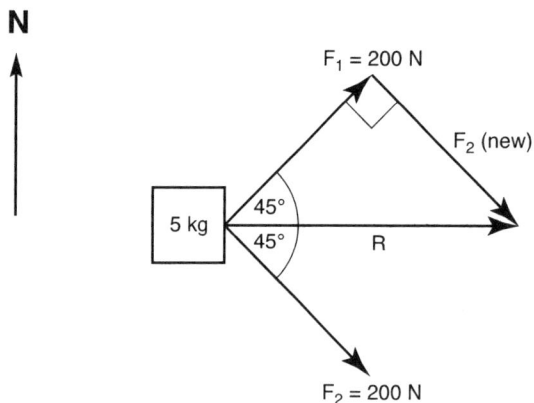

Since F_1 and F_2 are equal (200 N) and the two angles are equal (45°), the diagram is symmetrical and the direction of the resultant force is east.

Moving F_2 to its new position creates a right-angled triangle.

(a)   Resultant force

$$R^2 = F_1{}^2 + F_2{}^2$$
$$R^2 = 200^2 + 200^2$$
$$R^2 = 40\,000 + 40\,000$$
$$R = \sqrt{80\,000}$$
$$R = 282{\cdot}84 \text{ N east (bearing } 090°)$$

(b)   Acceleration

$$a = \frac{F}{m}$$
$$a = \frac{282{\cdot}84}{5}$$
$$a = 56{\cdot}57 \text{ m s}^{-2} \text{ east (bearing } 090°)$$

EXAMPLE 2

Problem

Two forces of 80 N pull a 7 kg mass (see diagram below).

A frictional force of 30 N acts against the direction of motion.

Find:

(a) the resultant force
(b) the acceleration of the mass.

Solution

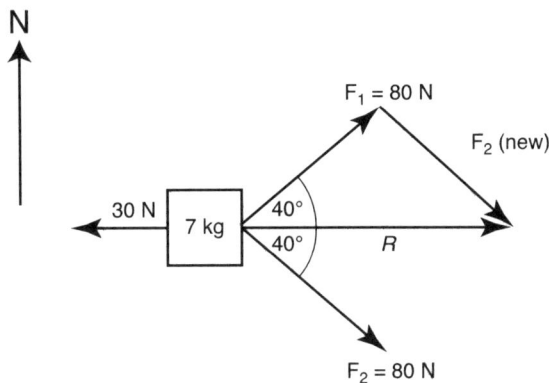

As before, resultant of F_1 and F_2 is east.

Moving F_2 to its new position creates an isosceles triangle.

(a) Resultant force

from sine rule: $\dfrac{R}{\sin 100} = \dfrac{80}{\sin 40}$

$R = \dfrac{80 \sin 100}{\sin 40}$

$R = \underline{122 \cdot 57 \text{ N east}}$

Now subtract the frictional force.

Resultant force $= 122 \cdot 57 - 30$

Resultant force $= \underline{92 \cdot 57 \text{ N east (bearing } 090°)}$

(b) Acceleration

$a = \dfrac{F}{m}$

$a = \dfrac{92 \cdot 57}{7}$

$a = \underline{13 \cdot 22 \text{ m s}^{-2} \text{ east (bearing } 090°)}$

Progress to P.P. 'H' P. Page 9, nos 13–18

Resolution of vectors

This is the opposite of addition of vectors where we resolve or separate a single vector into its component vectors. Very often this is referred to as resolution of forces since we normally wish to find the two component forces (*x* and *y*) of a resultant force R. We consider only the simple case where *x* and *y* are perpendicular to each other.

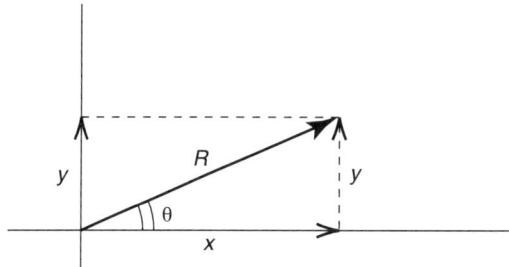

From the diagram:

$$\sin \theta = \frac{y}{R} \qquad\qquad\qquad \cos \theta = \frac{x}{R}$$

$$y = \underline{\underline{R \sin \theta}} \qquad\qquad\qquad x = \underline{\underline{R \cos \theta}}$$

These two formulae give the magnitudes of the component vectors *x* and *y*.

Resolving the weight of a mass on an incline

Suppose a 5 kg mass slides down a 30° slope. Its weight, i.e., the force of gravity, acts vertically downwards.

$$\begin{aligned}
\text{weight} &= m\,g \\
&= 5 \text{ kg} \times 9{\cdot}8 \text{ N kg}^{-1} \\
&= \underline{\underline{49 \text{ N}}}
\end{aligned}$$

This gravitational force of 49 N can be resolved into two components, one parallel to the surface and one perpendicular to the surface (see diagram below).

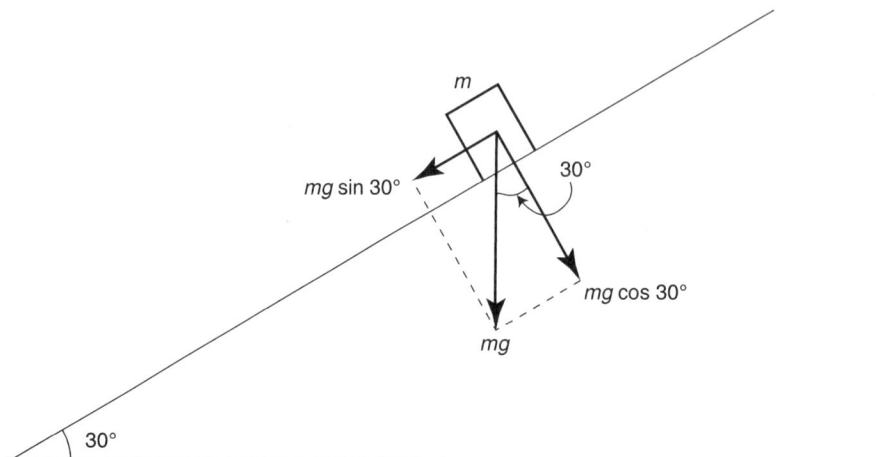

From the diagram:

$$\sin 30° = \frac{\text{parallel component}}{m\,g} \qquad\qquad \cos 30° = \frac{\text{perpendicular component}}{m\,g}$$

parallel components = m g sin 30°   perpendicular components = m g cos 30°

As before, these two formulae give the magnitudes of the component vectors.

The parallel component   = m g sin 30°
                                = 49 sin 30°
                                = 24·5 N

The perpendicular component   = m g cos 30°
                                = 49 cos 30°
                                = 42·44 N

The normal reaction (N.R.) is a force which exactly balances the perpendicular component, i.e., it is the force exerted by the surface on the mass.

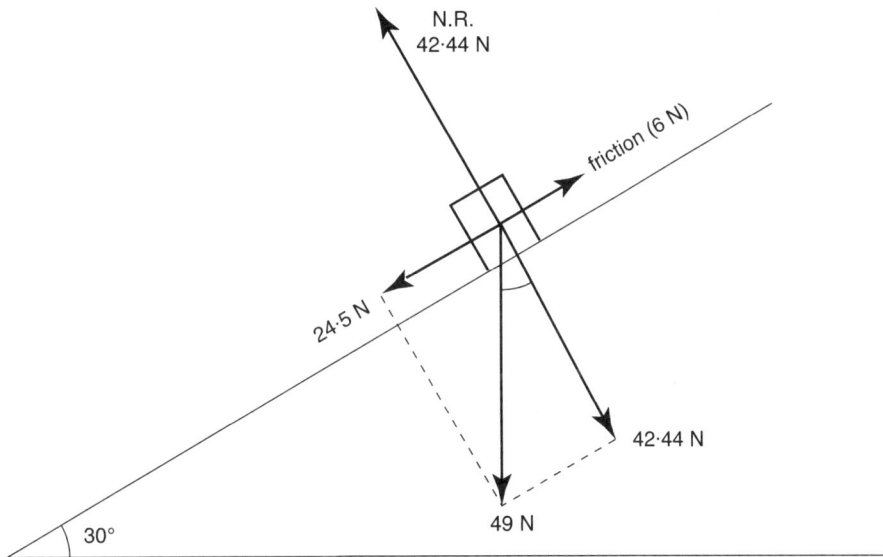

Friction, too, has to be taken into account and the frictional force acts against the parallel component, i.e., against the direction of the accelerating force. Therefore, the force causing the 5 kg mass to accelerate is (24·5 N − frictional force) and consequently the acceleration of the

5 kg mass is $\left(\dfrac{24\cdot5\ \text{N} - \text{frictional force}}{5\ \text{kg}} \right)$.

$$a = \frac{24\cdot5 - 6}{5}$$

$$a = 3\cdot7\ \text{m s}^{-2}$$

Progress to P.P. 'H' P. Page 10, nos 19–22

CHAPTER 2

EQUATIONS OF MOTION

GRAPHS OF MOTION

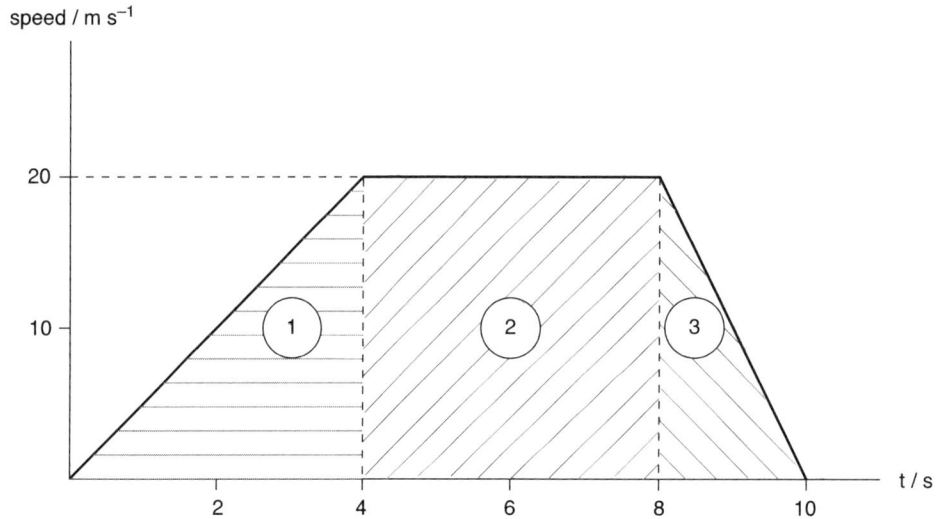

---

1. Acceleration (first 4 s)

$$a = \frac{v - u}{t}$$

$$a = \frac{20 - 0}{4}$$

$$a = \underline{\underline{5 \text{ m s}^{-2}}}$$

---

2. Deceleration (last 2 s)

$$a = \frac{v - u}{t}$$

$$a = \frac{0 - 20}{2}$$

$$a = \underline{\underline{-10 \text{ m s}^{-2}}}$$

$$\therefore \quad \text{deceleration} = \underline{\underline{10 \text{ m s}^{-2}}}$$

SPEED / TIME GRAPHS

The graph opposite shows the motion of a car.

The car accelerates constantly from rest to 20 m s^{-1} in 4 s.

Next the car travels at constant speed for another 4 s.

Finally the car decelerates to rest in 2 s.

Four different quantities can be calculated from the graph.

3. Total distance travelled total distance travelled = area under graph Area ① $= \frac{1}{2} \times 4 \times 20 = $ 40 m Area ② $= \quad 4 \times 20 = $ 80 m Area ③ $= \frac{1}{2} \times 2 \times 20 = $ 20 m Total distance travelled $=$ 140 m	**4.** Average speed average speed $= \dfrac{\text{total distance}}{\text{total time}}$ $= \dfrac{140}{10}$ $= $ 14 m s^{-1}

LEARN
Total distance travelled $\quad = $ area under a $\quad\quad\quad$ speed / time graph

Progress to P.P. 'H' P. Page 12, nos 1–3

Plotting an acceleration / time graph from a velocity / time graph

Given the following velocity / time graph, acceleration can be calculated for each section.

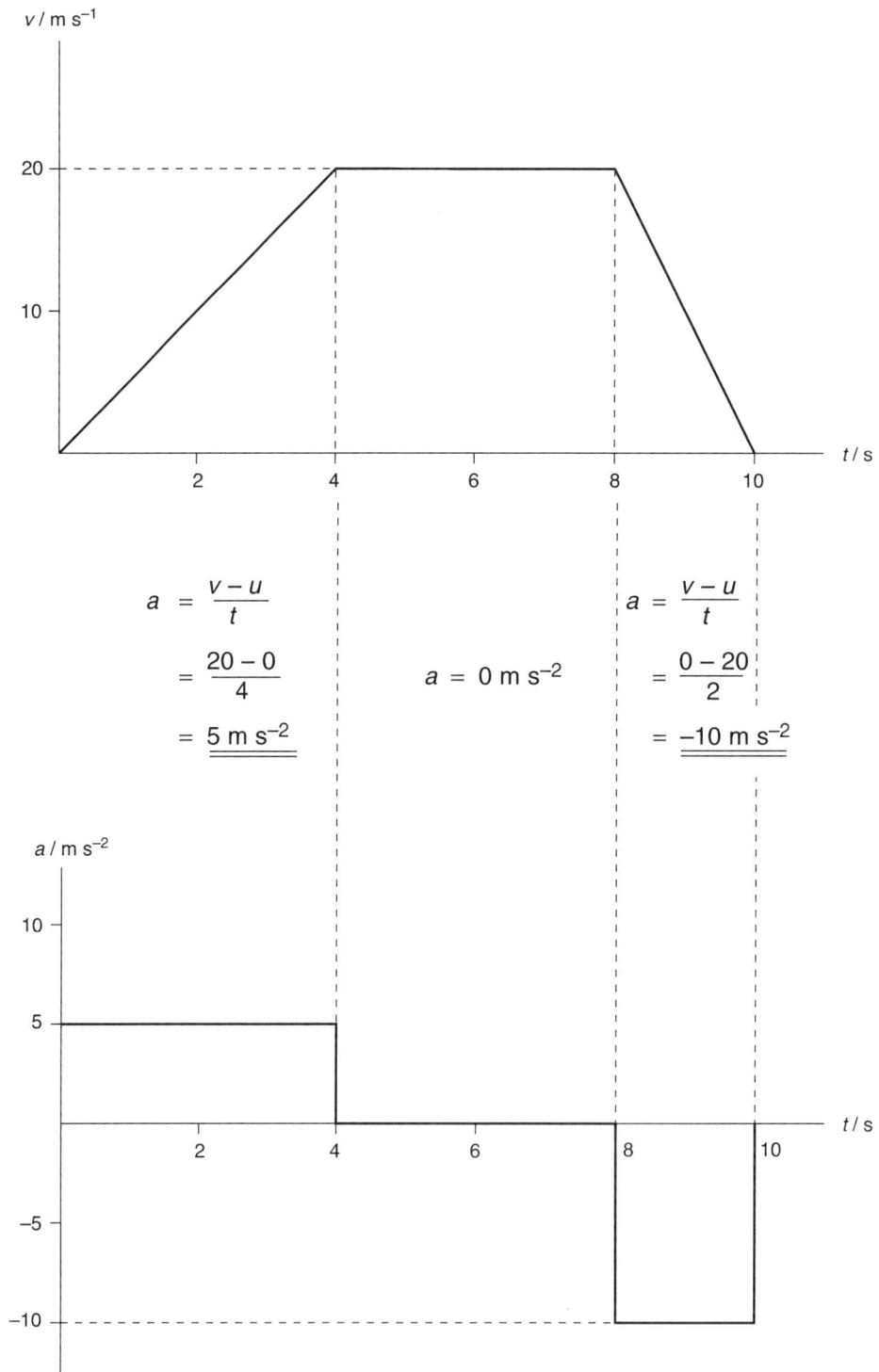

$$a = \frac{v - u}{t}$$

$$= \frac{20 - 0}{4}$$

$$= \underline{\underline{5 \text{ m s}^{-2}}}$$

$$a = 0 \text{ m s}^{-2}$$

$$a = \frac{v - u}{t}$$

$$= \frac{0 - 20}{2}$$

$$= \underline{\underline{-10 \text{ m s}^{-2}}}$$

The gradient of the velocity / time graph gives acceleration.

Comparing a speed / time graph with a velocity / time graph

A ball is thrown vertically up at 19·6 m s^{-1}.
It takes 2 seconds to go up and 2 seconds to come down.
The first graph is a speed / time graph.
The second graph is a velocity / time graph.

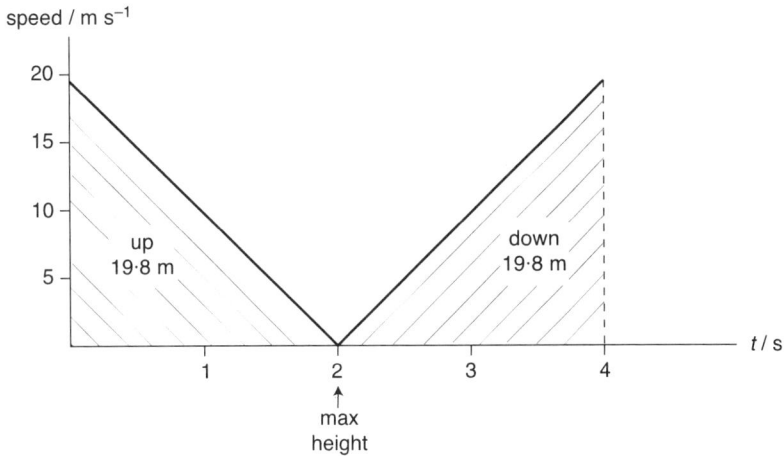

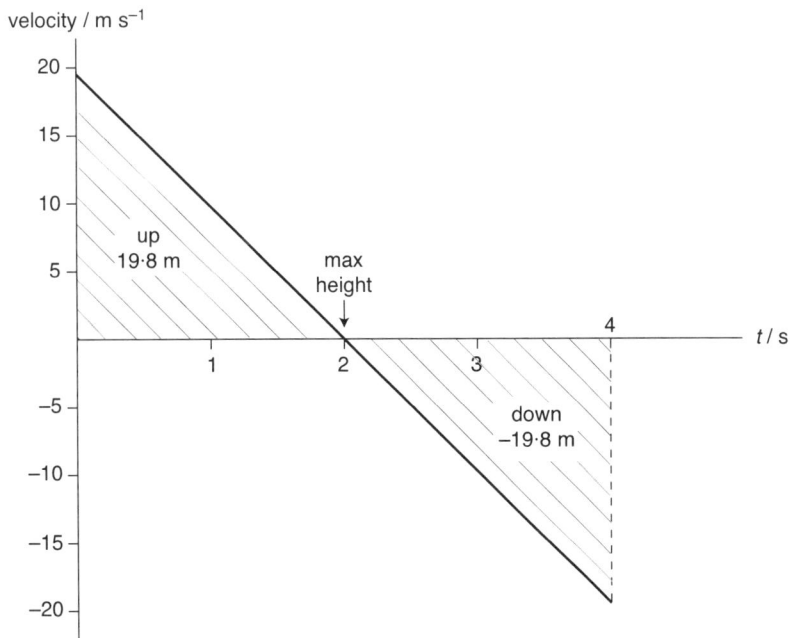

Both graphs show the same motion.
Speed is a scalar quantity (always positive).
Velocity is a vector quantity (hence the negative y-axis).
Total distance gone = 39·6 m
Total displacement = 0 m

Progress to P.P. 'H' P. Page 12, nos 4–8

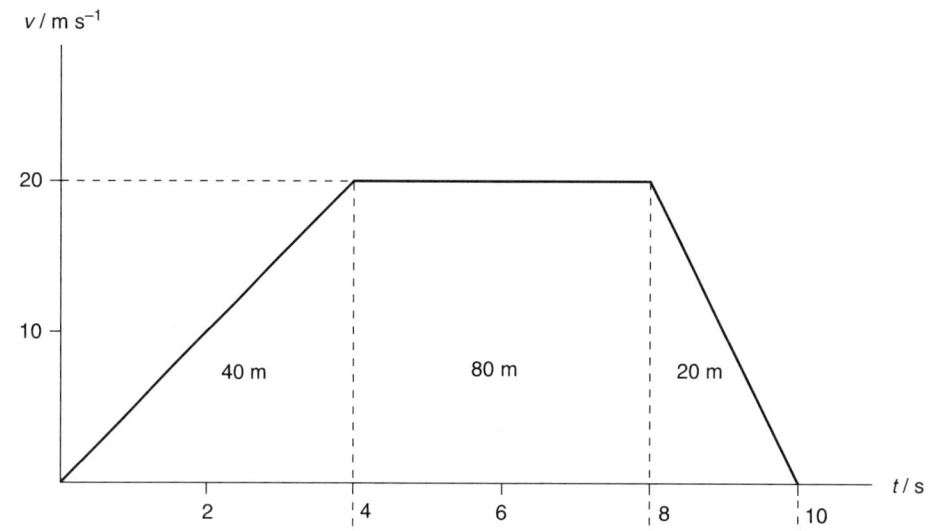

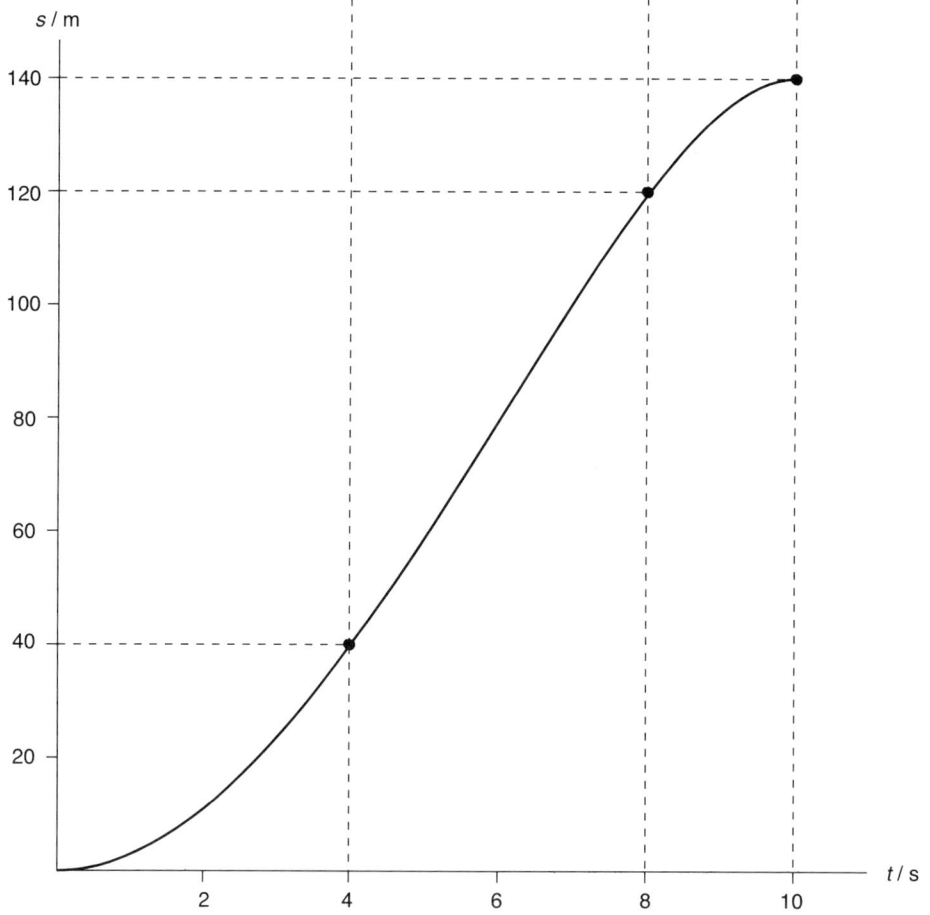

Plotting a displacement / time graph from a velocity / time graph

Given the velocity / time graph opposite, displacement can be calculated from the area under the graph.

First 4 seconds.

time / s	displacement / m
0	0
1	$\frac{1}{2} \times 1 \times 5 = 2 \cdot 5$
2	$\frac{1}{2} \times 2 \times 10 = 10$
3	$\frac{1}{2} \times 3 \times 15 = 22 \cdot 5$
4	$\frac{1}{2} \times 4 \times 20 = 40$

Next 4 seconds.

time / s	displacement / m
5	$1 \times 20 = 20 \quad | \quad + 40 = 60$
6	$2 \times 20 = 40 \quad | \quad + 40 = 80$
7	$3 \times 20 = 60 \quad | \quad + 40 = 100$
8	$4 \times 20 = 80 \quad | \quad + 40 = 120$

Last 2 seconds.

time / s	displacement / m
9	$140 \text{ (total)} - 5 = 135$
10	$40 + 80 + 20 = 140$

The displacement / time graph opposite is constructed from these points.

Progress to P.P. 'H' P. Page 14, nos 9–12

Calculating Acceleration Experimentally : Method 1

A trolley with a 10 cm card mounted on it is released from rest and rolls down an incline. The card cuts two light beams, each of which is connected to a separate electronic timer. A stopwatch measures the time taken for the trolley to travel from X to Y.

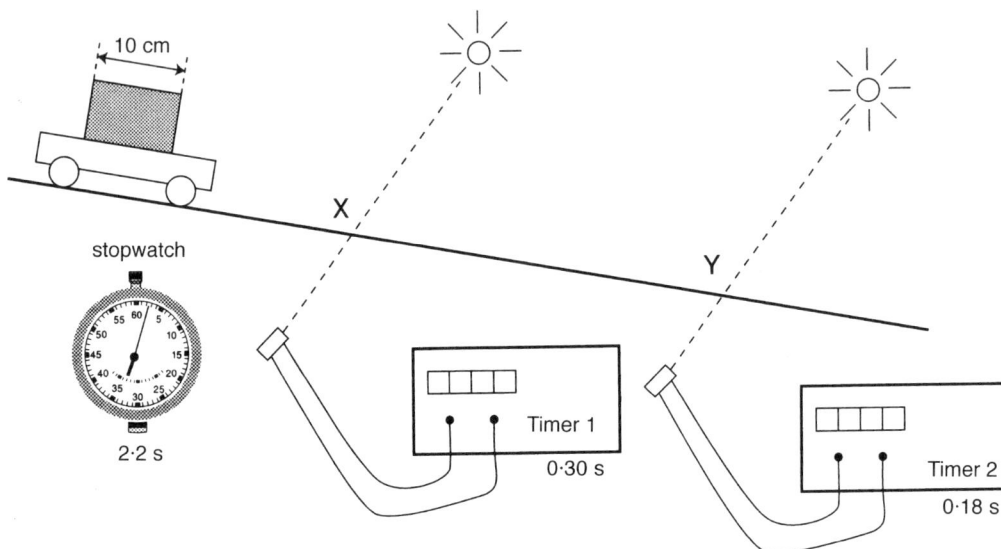

The experimental results are shown in the diagram.

Before using the acceleration equation, it is necessary to find the three quantities.

$$u = \frac{\text{length of card}}{\text{time on timer 1}} = \frac{0.10 \text{ m}}{0.30 \text{ s}} = 0.33 \text{ m s}^{-1}$$

$$v = \frac{\text{length of card}}{\text{time on timer 2}} = \frac{0.10 \text{ m}}{0.18 \text{ s}} = 0.55 \text{ m s}^{-1}$$

$$t = \text{time on stopwatch} = 2.2 \text{ s}$$

Now use the equation

$$a = \frac{v - u}{t}$$

$$a = \frac{0.55 - 0.33}{2.2}$$

$$a = \underline{\underline{0.10 \text{ m s}^{-2}}}$$

Calculating Acceleration Experimentally : Method 2

Method 2 is the same as method 1 only this time there is no stopwatch. Instead a meter stick is used to measure the distance between X and Y.

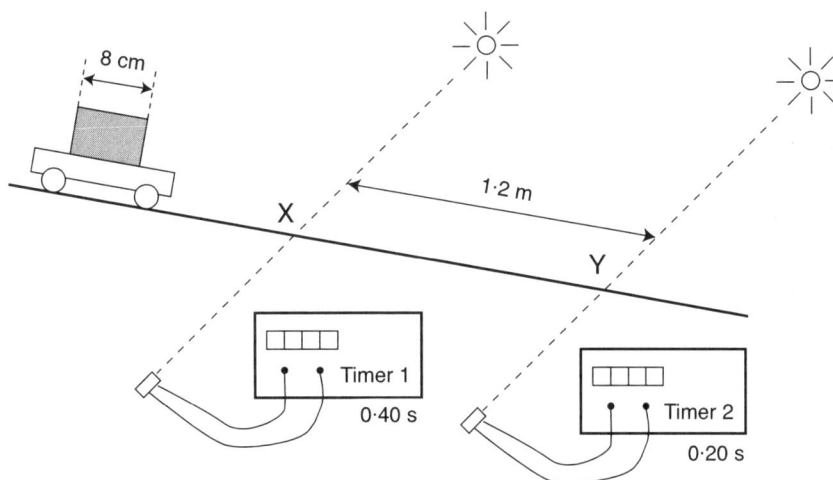

The experimental results are shown in the diagram.

First: calculate u and v

$$u = \frac{\text{length of card}}{\text{time on timer 1}} = \frac{0 \cdot 08 \text{ m}}{0 \cdot 40 \text{ s}} = 0 \cdot 20 \text{ m s}^{-1}$$

$$v = \frac{\text{length of card}}{\text{time on timer 2}} = \frac{0 \cdot 08 \text{ m}}{0 \cdot 20 \text{ s}} = 0 \cdot 40 \text{ m s}^{-1}$$

Second: find the average velocity

$$\bar{v} = \frac{u + v}{2} = 0 \cdot 30 \text{ m s}^{-1}$$

Third: find the time between X and Y:

$$t = \frac{d}{\bar{v}} = \frac{1 \cdot 2 \text{ m}}{0 \cdot 30 \text{ m s}^{-1}} = 4 \text{ s}$$

Fourth: now use the acceleration equation.

$$a = \frac{v - u}{t} = \frac{0 \cdot 40 - 0 \cdot 20}{4} = \underline{\underline{0 \cdot 05 \text{ m s}^{-2}}}$$

N.B. With a knowledge of equations of motion the student could use $v^2 = u^2 + 2as$

LEARN
$a = \dfrac{v - u}{t}$
average speed $= \dfrac{u + v}{2}$

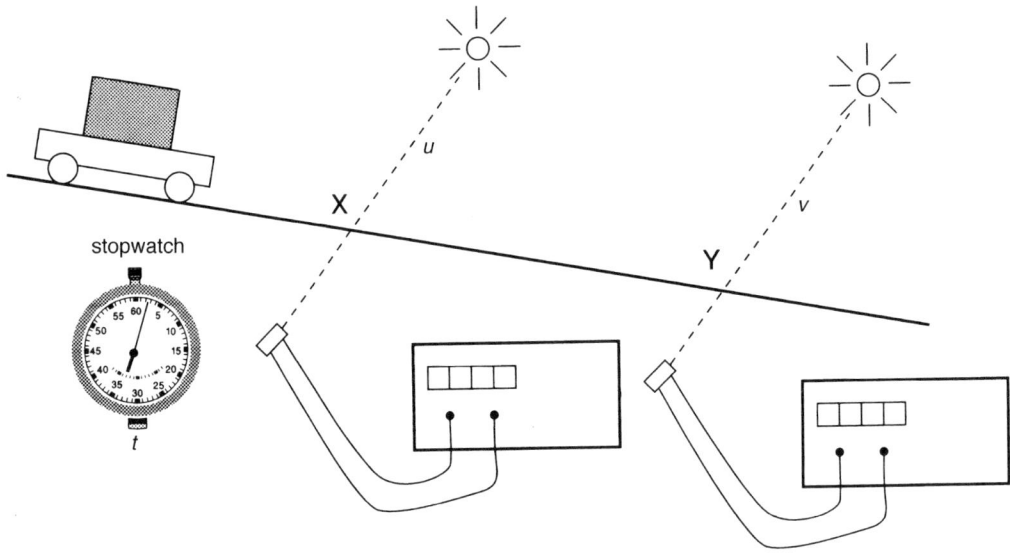

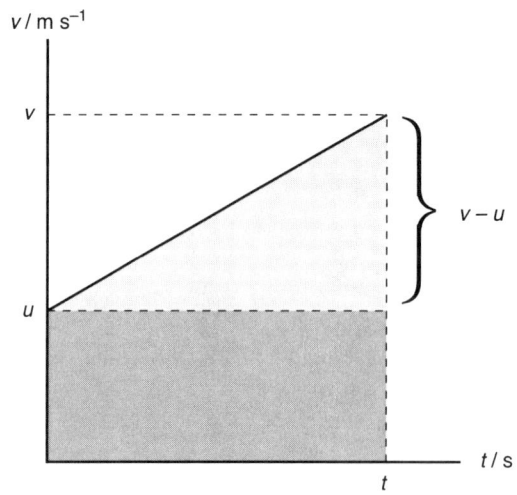

THE EQUATIONS OF MOTION

The First Equation of Motion

A trolley with a card mounted on it is released from rest and rolls down an incline (opposite).
The card cuts two light beams, each of which is connected to a separate electronic timer.
A stopwatch measures the time taken for the trolley to travel from X to Y.

The acceleration can be calculated from the equation.

$$a \quad = \frac{v - u}{t}$$

$$\therefore \qquad at \quad = v - u$$

$$\therefore \quad u + at \quad = v$$

$$\therefore \qquad v \quad = \underline{\underline{u + at}}$$

$$\boxed{v \quad = \quad u + at}$$

The Second Equation of Motion

The trolley in the diagram (top left opposite) accelerates constantly from u to v in time t. The velocity / time graph opposite shows the motion of this trolley.

The total distance gone $=$ area under the graph

$$s \quad = \quad \text{area of rectangle } + \text{ area of triangle}$$

$$s \quad = \quad ut + \frac{1}{2}t\,(v - u)$$

$$s \quad = \quad ut + \frac{1}{2}t \times at \text{ [since } v - u = at]$$

$$s \quad = \quad ut + \frac{1}{2}at^2$$

$$\boxed{s \quad = \quad ut + \frac{1}{2}at^2}$$

LEARN
$v = u + at$
$s = ut + \dfrac{1}{2}at^2$

Worked Example on the First Equation of Motion

Problem

A car accelerates constantly at 2 m s^{-2}. How long does it take to accelerate from 3 m s^{-1} to 5·5 m s^{-1}?

Solution

Always start with a list of five quantities, filling in those which are known.

$u = 3$ $v = u + at$

$v = 5·5$ $5·5 = 3 + 2t$

$a = 2$ $2·5 = 2t$

$s = /$

$t = ?$ $\therefore \ t = \dfrac{2·5}{2}$

$t = \underline{\underline{1·25 \ s}}$

Worked Example on the first two Equations of Motion

Problem

A ball falling under the influence of gravity passes two points, A and B. The velocity at A is 3 m s^{-1} and the velocity at B is 12 m s^{-1}. Calculate the distance between A and B.

A

$\downarrow$ 3 m s^{-1}

B

$\downarrow$ 12 m s^{-1}

Solution

Start with the list as before.

$u = 3$

$v = 12$

$a = 9·8$

$s = ?$

$t = /$

It is not possible to use

$s = ut + \dfrac{1}{2}at^2$ without

first finding t.

Step 1 $v = u + at$

$12 = 3 + 9·8t$

$9 = 9·8t$

$t = \dfrac{9}{9·8}$

$t = \underline{\underline{0·92 \ s}}$

Step 2 $s = ut + \dfrac{1}{2}at^2$

$s = 3(0·92) + \dfrac{1}{2} \ 9·8(0·92)^2$

$s = 2·76 + 4·147$

$s = \underline{\underline{6·91 \ m}}$

Notice that it required two equations to obtain the answer

Progress to P.P. 'H' P. Page 15, nos 1–8

The Third Equation of Motion

Sometimes it is convenient to have an equation which does not depend on t. The first and second equations can be combined to form such an equation.

$$v = u + at \qquad ① $$
$$s = ut + \frac{1}{2}at^2 \qquad ② $$

Square equation ①:

$$v^2 = (u + at)^2$$
$$v^2 = u^2 + 2uat + a^2t^2$$

$$v^2 = u^2 + 2a\left(ut + \frac{1}{2}at^2\right)$$

$$v^2 = \underline{\underline{u^2 + 2as}} \qquad \left[\text{since } s = ut + \frac{1}{2}at^2\right]$$

$$\boxed{v^2 = u^2 + 2as}$$

Worked Example on the Third Equation of Motion

The problem opposite, requiring the first two equations of motion, can be solved with a single equation.

List

$$u = 3$$
$$v = 12$$
$$a = 9.8$$
$$s = ?$$
$$t = /$$

$$v^2 = u^2 + 2as$$
$$12^2 = 3^2 + 2 \times 9.8\,s$$
$$144 - 9 = 19.6\,s$$
$$s = \frac{135}{19.6}$$
$$s = \underline{6.89\ m}$$

The difference between this answer and the 6·91 m opposite is due to earlier rounding up in step 1 opposite.

The 6·89 m answer is more accurate.

LEARN
$v^2 = u^2 + 2as$

Progress to P.P. 'H' P. Page 15, nos 9–12

Worked Example involving two directions

Problem

A helicopter is travelling upward at a steady velocity of 5 m s^{-1} when a parcel falls out of the door. 4·2 s later the parcel hits the ground.

(a)   How high was the helicopter when the parcel fell out?

(b)   How high was the helicopter when the parcel hit the ground?

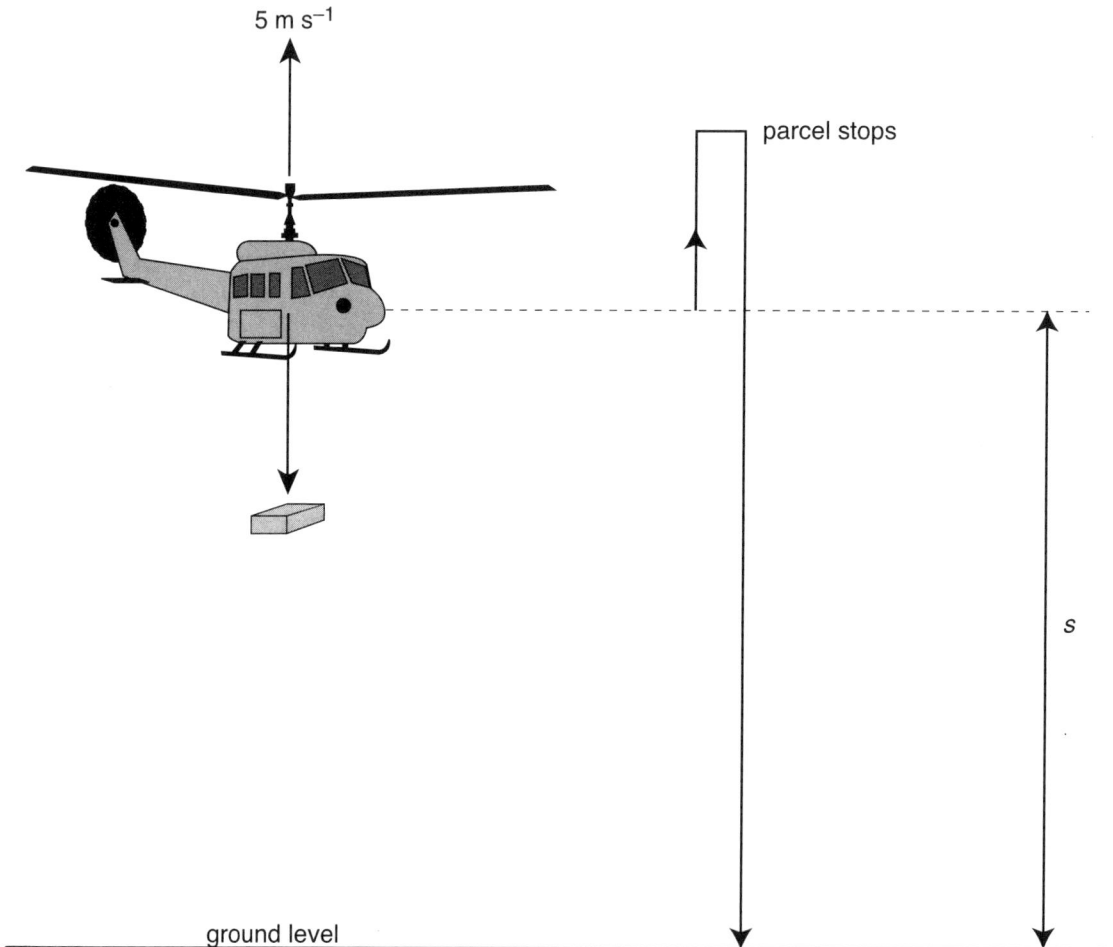

5 m s^{-1}

parcel stops

ground level

s

Solution

Initially the parcel moves upwards until it stops and then falls to ground level.

(a) The "trick" in this style of question is to realise that there are two directions involved.

To say the initial velocity of the parcel is +5 m s^{-1} means that upward direction is positive.

Therefore downward direction must be negative and $a = -9 \cdot 8$ m s^{-2}. (The choice of direction is arbitrary.)

List
$$s = ut + \frac{1}{2}at^2$$

$u = +5$
$$s = 5 \times 4 \cdot 2 + \frac{1}{2}(-9 \cdot 8)(4 \cdot 2)^2$$

$v = /$

$a = -9 \cdot 8$
$$s = 21 - 86 \cdot 436$$

$s = ?$

$t = 4 \cdot 2$
$$s = \underline{-65 \cdot 44 \text{ m}}$$

I.e., the helicopter was <u>65·44 m above the ground</u> when the parcel fell out.

(b) While the parcel is falling the helicopter is moving upwards at a constant speed of 5 m s^{-1}. Therefore the helicopter has travelled at 5 m s^{-1} for 4·2 s.

Distance gone by helicopter = speed × time
= 5 × 4·2
= <u>21 m</u>

But helicopter was already 65·44 m above the ground
∴ height of helicopter is now 21 + 65·44
= <u>86·44 m</u> when the parcel hit the ground.

Progress to P.P. 'H' P. Page 16, nos 1–8

PROJECTILE MOTION

Constant Vertical Acceleration

When a body falls under the influence of gravity, it accelerates constantly at 9·8 m s^{-2}.

Take the example of a ball dropped from a cliff.

After 1 second   the velocity = 9·8 m s^{-1}
After 2 seconds the velocity = 19·6 m s^{-1}
After 3 seconds the velocity = 29·4 m s^{-1} and the ball hits the ground.

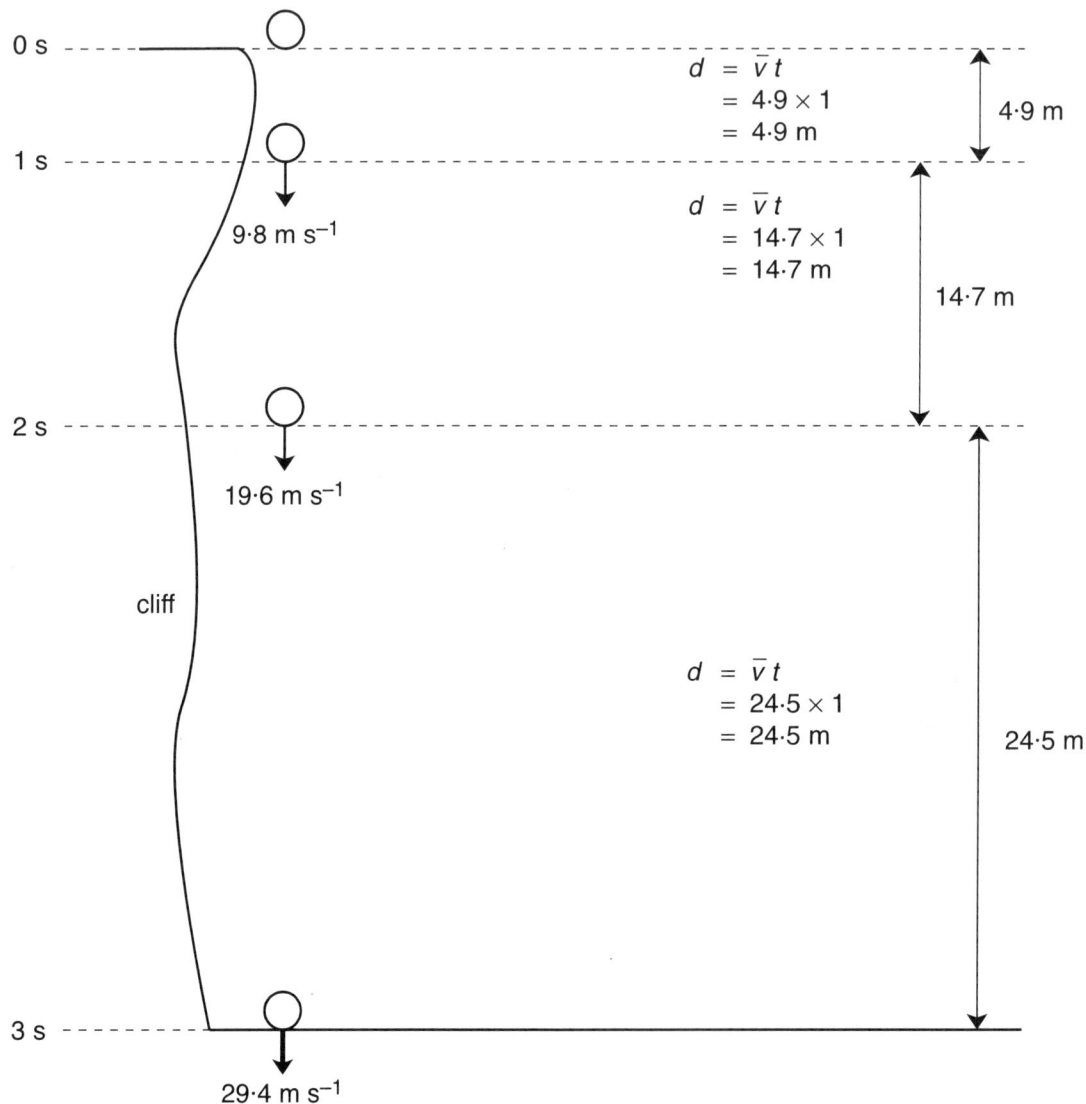

0 s

$d = \bar{v} t$
$= 4·9 \times 1$
$= 4·9$ m

4·9 m

1 s

9·8 m s^{-1}

$d = \bar{v} t$
$= 14·7 \times 1$
$= 14·7$ m

14·7 m

2 s

19·6 m s^{-1}

cliff

$d = \bar{v} t$
$= 24·5 \times 1$
$= 24·5$ m

24·5 m

3 s

29·4 m s^{-1}

The diagram also shows the distance dropped in each second.

How high is the cliff?

Obviously 4·9 + 14·7 + 25·5 = <u>44·1 m</u>

or $s = ut + \frac{1}{2}at^2$

$= \frac{1}{2} 9·8(3)^2$

$= \underline{44·1 \text{ m}}$

Constant Horizontal Velocity and Constant Vertical Acceleration

Imagine that when the ball (on the previous page) was dropped, a second ball was kicked horizontally at 7 m s⁻¹.

The kicked ball has the same **vertical acceleration** as the dropped ball but **horizontally** it travels at **constant velocity** (7 m s⁻¹).

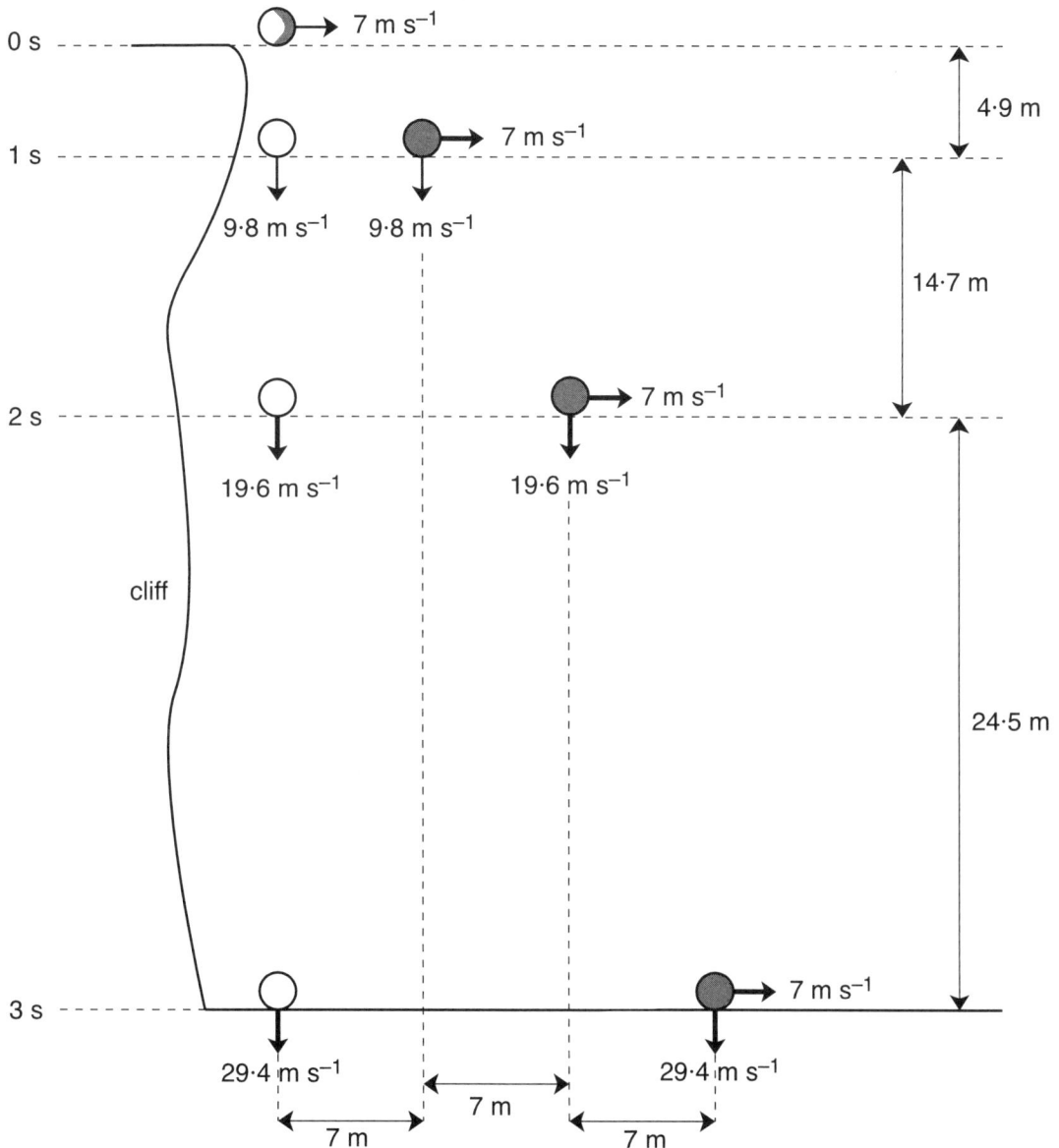

Both balls strike the ground at the same time but they are 21 m apart.

LEARN
Vertical motion is constant acceleration.
Horizontal motion is constant velocity.

Addition of velocities to find resultant velocity

At the end of the 1st second:

The kicked ball has a horizontal velocity of 7 m s^{-1} and a vertical velocity of 9·8 m s^{-1} .

The velocity at the end of the 1st second is the resultant of these two velocities.

$$R^2 = 7^2 + 9\cdot 8^2$$
$$R^2 = 49 + 96\cdot 04$$
$$R = \sqrt{145\cdot 04}$$
$$R = 12\cdot 04 \text{ m s}^{-1}$$

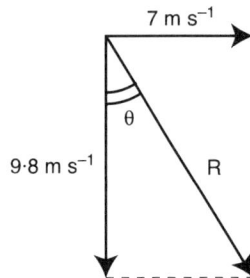

$$\tan \theta = \frac{7}{9\cdot 8}$$
$$\tan \theta = 0\cdot 714$$
invert tan
$$\therefore \quad \theta = 35\cdot 5°$$

$$R = 12\cdot 04 \text{ m s}^{-1} \quad 35\cdot 5° \text{ to the vertical}$$

At the end of the 2nd second:

The kicked ball has a horizontal velocity of 7 m s^{-1} and a vertical velocity of 19·6 m s^{-1} .

The velocity at the end of the 2nd second is the resultant of these two velocities.

$$R^2 = 7^2 + 19\cdot 6^2$$
$$R^2 = 49 + 384\cdot 16$$
$$R = \sqrt{433\cdot 16}$$
$$R = 20\cdot 81 \text{ m s}^{-1}$$

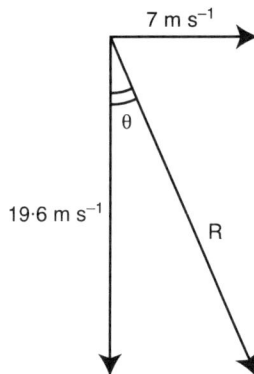

$$\tan \theta = \frac{7}{19\cdot 6}$$
$$\tan \theta = 0\cdot 357$$
invert tan
$$\therefore \quad \theta = 19\cdot 7°$$

$$R = 20\cdot 81 \text{ m s}^{-1} \quad 19\cdot 7° \text{ to the vertical}$$

At the end of the 3rd second:

The kicked ball has a horizontal velocity of 7 m s^{-1} and a vertical velocity of 29·4 m s^{-1} .

The velocity at the end of the 3rd second is the resultant of these two velocities.

R^2 = $7^2 + 29.4^2$

R^2 = 49 + 864·36

R = $\sqrt{913.36}$

R = 30·22 m s^{-1}

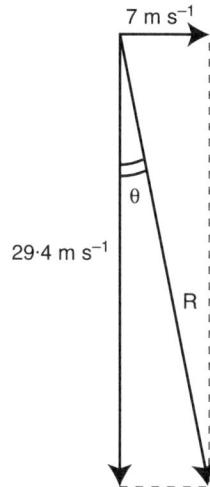

$\tan \theta = \dfrac{7}{29.4}$

$\tan \theta = 0.238$

invert tan

$\therefore \; \theta = 13.4°$

R = 30·22 m s^{-1} 13·4° to the vertical

Progress to P.P. 'H' P. Page 18, nos 1–9

Projectile fired at an angle to the horizontal

When a projectile, like an arrow, is fired with a velocity (40 m s^{-1}) at an angle (30°) to the horizontal, various quantities, like range, can be calculated.

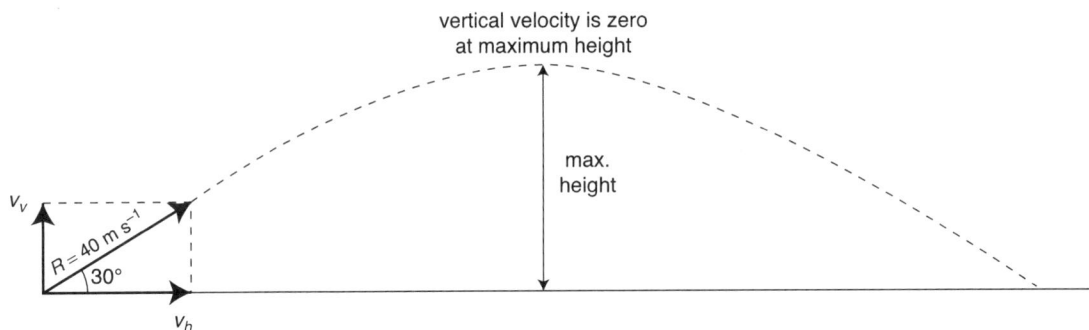

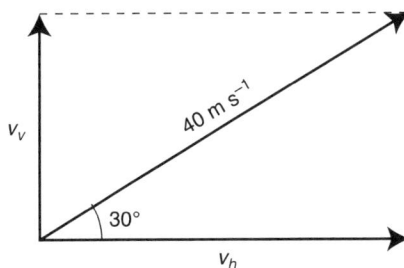

Horizontal Velocity:

$$\cos 30 = \frac{v_h}{40}$$

$$v_h = 40 \cos 30$$

$$v_h = \underline{\underline{34 \cdot 64 \text{ m s}^{-1}}}$$

Vertical Velocity:

$$\sin 30 = \frac{v_v}{40}$$

$$v_v = 40 \sin 30$$

$$v_v = \underline{\underline{20 \text{ m s}^{-1}}}$$

Time of flight:

Vertically (up)	$v = u + at$
$u = 20$	$0 = 20 + (-9{\cdot}8)t$
$v = 0$	$9{\cdot}8t = 20$
$a = -9{\cdot}8$	$t = \dfrac{20}{9{\cdot}8}$
$s = /$	
$t = ?$	$t = 2{\cdot}04$ s to go up

$$\therefore \ \text{time of flight} \ = \ \underline{\underline{4{\cdot}08 \text{ s}}}$$

Maximum height:

Vertically (up)

$u = 20$

$v = 0$

$a = -9{\cdot}8$

$s = ?$

$t = 2{\cdot}04$ (up)

$s = ut + \dfrac{1}{2}at^2$

$s = 20 \times 2{\cdot}04 + \dfrac{1}{2}(-9{\cdot}8)(2{\cdot}04)^2$

$s = 40{\cdot}8 - 20{\cdot}39$

$s = \underline{\underline{20{\cdot}41 \text{ m}}}$

Horizontal range:

horizontal velocity is constant

$\therefore$ horizontal distance $= v_h t$

range $= 34{\cdot}6 \times 4{\cdot}08$

range $= \underline{\underline{141{\cdot}17 \text{ m}}}$

Progress to P.P. 'H' P. Page 22, nos 10–16

CHAPTER 3

NEWTON'S SECOND LAW, ENERGY AND POWER

NEWTON'S SECOND LAW

Previously, from standard grade:

$$a \propto F \quad \text{[Provided mass is constant.]}$$

$$a \propto \frac{1}{m} \quad \text{[Provided force is constant.]}$$

Combining the relationships:

$$a \propto \frac{F}{m}$$

$$\therefore \quad a = \frac{kF}{m} \quad \text{where } k \text{ is a constant}$$

1 Newton is **defined** as the unbalanced force required to cause a 1 kg mass to accelerate at 1 m s^{-2}.

$$\therefore \quad k = 1$$

$$\therefore \quad a = \frac{F}{m}$$

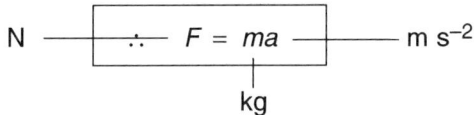

N ———| ∴ $F = ma$ |——— m s^{-2}

kg

EXAMPLE 1

Problem

Calculate the acceleration of the trolley shown.

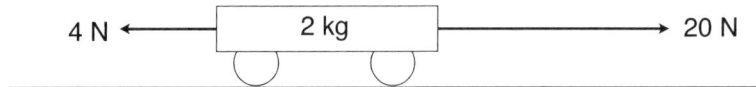

4 N ←——— | 2 kg | ———→ 20 N

Solution

The unbalanced force is 20 N – 4 N $=$ 16 N to the right

The acceleration is $\quad a = \dfrac{F}{m}$

$$a = \frac{16}{2}$$

$$a = 8 \text{ m s}^{-2} \text{ to the right.}$$

EXAMPLE 2

Problem

A 3 kg mass and a 9 kg mass are linked by a string passed over a frictionless pulley.

Calculate the acceleration of the system.

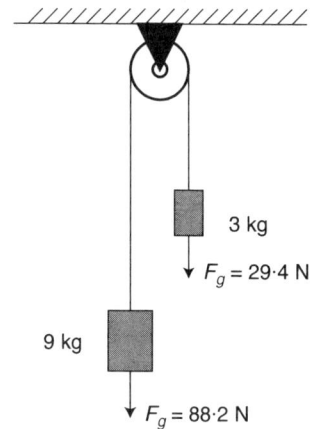

3 kg

$F_g = 29.4$ N

Solution

The unbalanced force causing the acceleration
$= 88.2 - 29.4$
$= \underline{\underline{58.8 \text{ N}}}$

This force is acting on a total mass of 12 kg.

$\therefore \quad a = \dfrac{F}{m} = \dfrac{58.8}{12} = \underline{\underline{4.9 \text{ m s}^{-2} \text{ (anticlockwise)}}}$

9 kg

$F_g = 88.2$ N

EXAMPLE 3

Problem

A 3 kg mass and a 9 kg mass are linked by a light string passed over a frictionless pulley. If the 9 kg mass is on a frictionless surface, find:

(a)   the acceleration of the system;
(b)   the tension in the string (T).

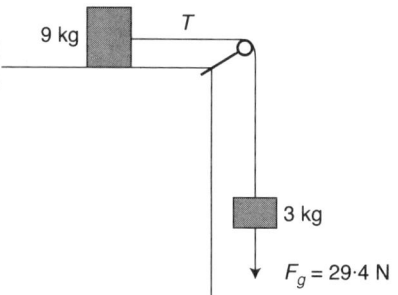

9 kg T

3 kg

$F_g = 29.4$ N

Solution

(a)   The unbalanced force is the force of gravity acting on the 3 kg mass.
$F_g = ma = 3 \times 9.8 = \underline{\underline{29.4 \text{ N}}}$

This force is pulling two masses, i.e., 12 kg.

$\therefore \quad a = \dfrac{F}{m} = \dfrac{29.4}{12} = \underline{\underline{2.45 \text{ m s}^{-2}}}$

(b)   The tension is the force pulling the 9 kg mass causing it to accelerate at 2.45 m s^{-2}.
$F_T = ma = 9 \times 2.45 = \underline{\underline{22.05 \text{ N}}}$

LEARN
$F = ma$

Progress to P.P. 'H' P. Page 25, nos 1–12

EXAMPLE 4

Problem

Three identical 0·3 kg trucks are connected by two light strings and pulled by a 1·8 N force by a toy train:

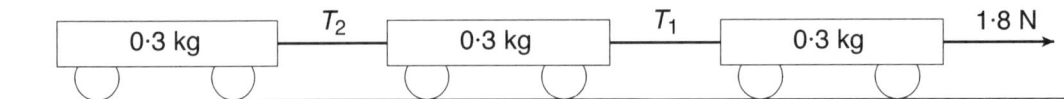

Assuming the surface to be frictionless,
(a) calculate the acceleration of the three trucks;
(b) find the tension in each string T_1 and T_2.

Solution

(a) The unbalanced force of 1·8 N is acting on a total mass of 0·9 kg.
The acceleration is:

$$a = \frac{F}{m} = \frac{1.8}{0.9} = \underline{\underline{2 \text{ m s}^{-2}}}$$

(b) T_1 is the force causing two trucks (0·6 kg) to accelerate at 2 m s^{-2}.

$$F_{T_1} = ma$$
$$T_1 = 0.6 \times 2$$
$$T_1 = \underline{\underline{1.2 \text{ N}}}$$

T_2 is the force causing one truck (0·3 kg) to accelerate at 2 m s^{-2}.

$$F_{T_2} = ma$$
$$T_2 = 0.3 \times 2$$
$$T_2 = \underline{\underline{0.6 \text{ N}}}$$

Progress to P.P. 'H' P. Page 28, nos 13–19

Students may wish to revise Resolution of Vectors in Chapter 1 before attempting Example 5.

EXAMPLE 5

Problem

A 3 kg trolley is placed on a 25° frictionless slope as shown. The force of gravity acting on the trolley is 29·4 N but only a component of this force (parallel to the slope), F_p causes the trolley to accelerate.

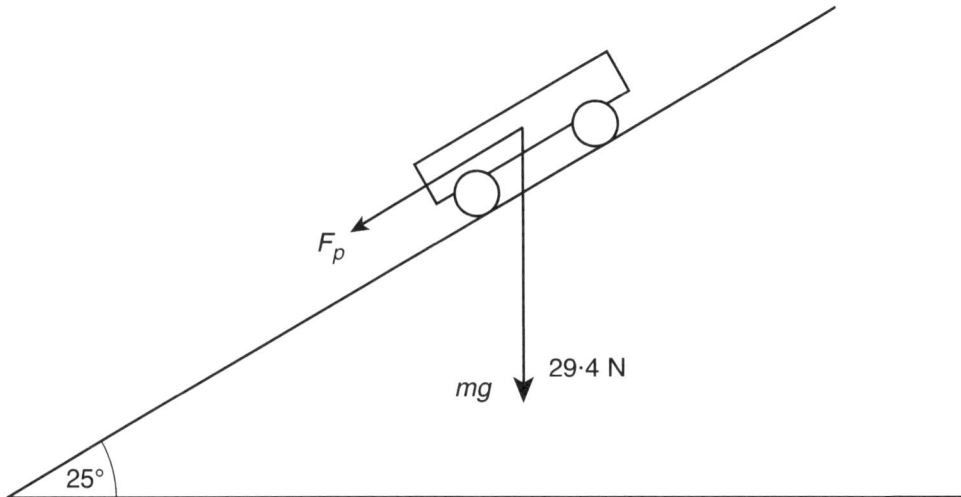

(a) Calculate the size of this accelerating force.

(b) Calculate the acceleration of the trolley (assuming friction is negligible).

Solution

(a) $\sin 25 = \dfrac{F_p}{mg}$

$F_p = mg \sin 25$

$F_p = 29{\cdot}4 \sin 25$

$F_p = \underline{\underline{12{\cdot}42 \text{ N}}}$

(b) $a = \dfrac{F}{m}$

$a = \dfrac{12{\cdot}42}{3}$

$a = \underline{\underline{4{\cdot}14 \text{ m s}^{-2}}}$

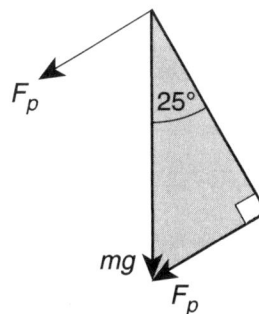

Progress to P.P. 'H' P. Page 30, nos 20–25

INTERCONVERSION OF POTENTIAL ENERGY AND KINETIC ENERGY

Previous equations from Standard Grade.

$$J \longrightarrow \boxed{E_p \quad = \quad mgh} \longrightarrow m$$
$$\text{kg} \qquad \text{N kg}^{-1}$$

$$J \longrightarrow \boxed{E_k \quad = \quad \tfrac{1}{2}mv^2} \longrightarrow \text{m s}^{-1}$$
$$\text{kg}$$

Potential energy can be converted into kinetic energy and vice versa.

EXAMPLE 1:   A falling body

Problem

When a 3 kg mass falls from A to B, its E_p is converted into E_k. Find its impact velocity.

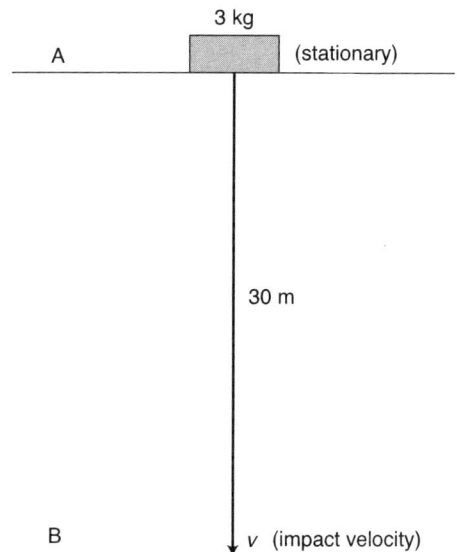

3 kg

A   (stationary)

30 m

B v (impact velocity)

Solution

Step 1: E_p at A
$$E_p = mgh$$
$$E_p = 3 \times 9{\cdot}8 \times 30$$
$$E_p = \underline{\underline{882 \text{ J}}}$$

As the mass falls its E_p is gradually converted into E_k.

Step 2: E_k at B
$$E_k = \underline{\underline{882 \text{ J}}}$$
i.e., no energy is lost.

Step 3:   Impact velocity

To calculate the impact velocity equate E_p at A with E_k at B.

E_k at B $= E_p$ at A

$$\tfrac{1}{2}mv^2 = 882$$
$$\tfrac{1}{2}3v^2 = 882$$
$$v^2 = \frac{882 \times 2}{3}$$
$$v = \sqrt{588}$$
$$v = \underline{\underline{24{\cdot}25 \text{ m s}^{-1}}}$$

Alternatively

$$\tfrac{1}{2}mv^2 = mgh$$
$$v^2 = 2gh$$
$$v = \sqrt{2gh}$$
$$v = \sqrt{2 \times 9{\cdot}8 \times 30}$$
$$v = \underline{\underline{24{\cdot}25 \text{ m s}^{-1}}}$$

EXAMPLE 2:   **The velocity of an air pellet**   [**N.B.** Knowledge of momentum required]

Problem

In a test designed to calculate the velocity of an air pellet, the pellet (mass 20 g) was fired horizontally at a lump of plasticine (mass 0·48 kg) suspended by a long string.

The plasticine moved up in height

$$h = 5 \text{ cm}$$

before swinging back.

Calculate the velocity of the air pellet before it hit the plasticine.

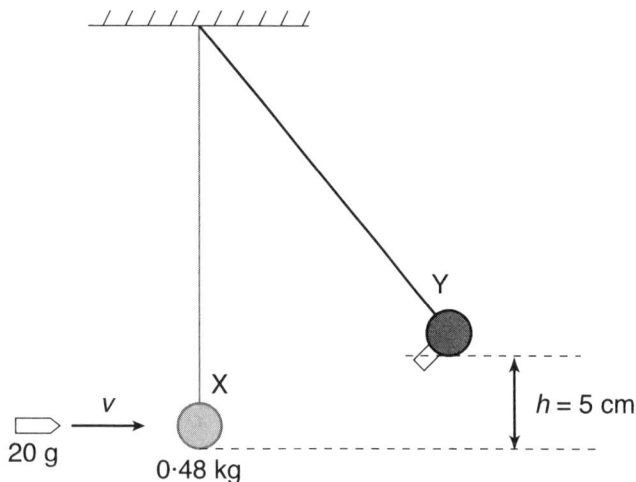

Solution

Step 1:   Calculate E_p at Y

$$E_p = mgh = 0.5 \times 9.8 \times 0.05 = \underline{0.245 \text{ J}}$$

Step 2:   This is the same as the E_k of pellet + plasticene at X = 0·245 J

Step 3:   Find velocity at X (after impact).

$$\frac{1}{2}mv^2 = 0.245$$

$$\frac{1}{2}0.5v^2 = 0.245$$

$$v^2 = 0.98$$

$$v = \underline{0.99 \text{ m s}^{-1}}$$

Step 4:   Find momentum at X (after impact).

momentum after $= mv$

$$= 0.5 \times 0.99$$

$$= \underline{0.495 \text{ kg m s}^{-1}}$$

Step 5:   This is the same as the momentum of the pellet before impact = 0·495 kg m s^{-1}

Step 6:   The velocity of the pellet before impact is:

$$v = \frac{\text{momentum}}{\text{mass of pellet}} = \frac{0.495}{0.02}$$

$$= \underline{\underline{24.75 \text{ m s}^{-1}}}$$

LEARN
$E_p = mgh$
$E_k = \frac{1}{2}mv^2$
$v = \sqrt{2gh}$

Progress to P.P. 'H' P. Page 33, nos 1–7, 8, 9

CHAPTER 4

MOMENTUM AND IMPULSE

MOMENTUM

Momentum is defined as the product of mass and velocity.

$$\text{kg m s}^{-1} \longrightarrow \boxed{p = mv} \longrightarrow \text{m s}^{-1}$$
$$\text{kg}$$

EXAMPLE:

Problem

How much momentum does a 45 kg boy have when he is running at 3 m s^{-1}?

Solution

$p = mv$
$p = 45 \times 3$
$p = 135$ kg m s^{-1}

Conservation of Momentum

The Law of Conservation of linear momentum applies to collisions and explosions between two objects moving in one dimension.

The total momentum before = the total momentum after
(collision or explosion)   (collision or explosion)
IN THE ABSENCE OF EXTERNAL FORCES

There are three categories of interaction:

A — Inelastic collisions — momentum is conserved — E_k is "lost"
B — Elastic collisions   — momentum is conserved — E_k is conserved
C — Explosions   — momentum is conserved — E_p is converted into E_k

Notice that momentum is always conserved.

MOMENTUM (A) — INELASTIC COLLISIONS

This is the "normal" everyday type of collision in which energy is lost (E_k is lost) and turned into other forms of energy.

EXAMPLE:

Problem

The example below shows how a 2 kg trolley travelling at 8 m s^{-1} collides with and sticks to an identical stationary trolley. Calculate
(a) velocity of both trolleys after collision;
(b) loss in kinetic energy.

Solution

(a)   Before   After

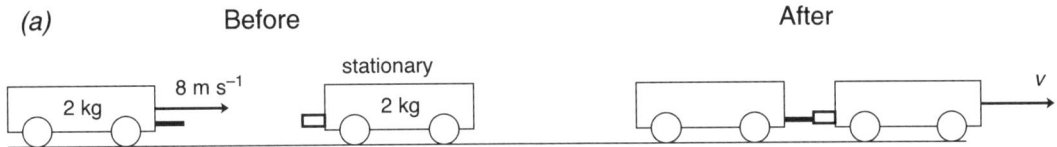

momentum before $= mv + mv$ momentum after $= mv$
$\qquad\qquad\quad = 2 \times 8 + 2 \times 0$ $16 \quad = 4v$
$\qquad\qquad\quad = 16$ kg m s^{-1} $\therefore \quad v \quad = 4$ m s^{-1}

This is the same as the total
momentum after

TOTAL MOMENTUM BEFORE = TOTAL MOMENTUM AFTER

(b) E_k before $= \frac{1}{2} mv^2 + \frac{1}{2} mv^2$ E_k after $= \frac{1}{2} mv^2$

$\qquad\qquad = \frac{1}{2} \times 2 \times (8)^2 + \frac{1}{2} \times 2 \times 0$ $= \frac{1}{2} \times 4 \times (4)^2$

$\qquad\qquad = 64$ J $= 32$ J

$\therefore$ 32 J of kinetic energy have been lost, i.e., converted into heat and sound in the collision.

Progress to P.P. 'H' P. Page 37, nos 1–22

LEARN
$p = mv$
$\begin{array}{cc} mv & = mv \\ \text{before} & \text{after} \end{array}$

MOMENTUM (B) — ELASTIC COLLISIONS

Momentum is always conserved in every type of collision or explosion.

In the **elastic** collision, both momentum and **kinetic energy** are **conserved**.

EXAMPLE:

Problem

On a linear air track, vehicle A (0·02 kg), moving at 1·0 m s^{-1}, collides with vehicle B (0·02 kg) which is initially stationary.

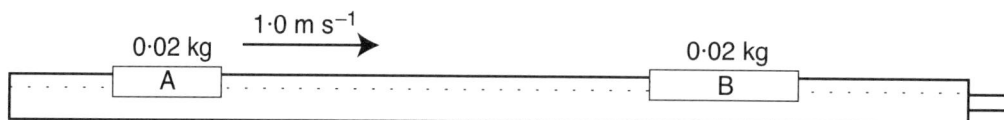

After the collision, vehicle A stops and vehicle B moves to the right.

(a) Find the momentum of A before the collision.

(b) Find the momentum of B after the collision.

(c) Hence calculate the speed of B after the collision.

(d) Is the collision elastic or inelastic? You must justify your answer with calculations.

Solution

(a) Momentum of A (before) $= mv = 0.02 \times 1 = \underline{\underline{0.02 \text{ kg m s}^{-1}}}$.

(b) Since A stops and momentum is conserved, the momentum of B (after) is the same as the momentum of A (before).

∴ momentum of B (after) $= \underline{\underline{0.02 \text{ kg m s}^{-1}}}$.

(c) $p = mv$

$0.02 = 0.02 \, v$

∴ $v = \underline{\underline{1 \text{ m s}^{-1}}}$

(d) E_k before $= \frac{1}{2}mv^2$ (A) E_k after $= \frac{1}{2}mv^2$ (B)

$= \frac{1}{2}0.02(1)^2$ $= \frac{1}{2}0.02\,(1)^2$

$= \underline{\underline{0.01 \text{ J}}}$ $= \underline{\underline{0.01 \text{ J}}}$

Therefore E_k is conserved.
Therefore the collision is **elastic**.

Progress to P.P. 'H' P. Page 42, nos 1–3

MOMENTUM (C) — EXPLOSIONS

Momentum is always conserved in every type of collision or explosion.

In an explosion the initial momentum is zero.

In an explosion the final (total) momentum is zero.

Energy is converted from potential energy (before explosion) to kinetic energy (after explosion).

EXAMPLE:

Problem:

A bullet of mass 50 g is fired from a 4 kg gun at 60 m s^{-1}.

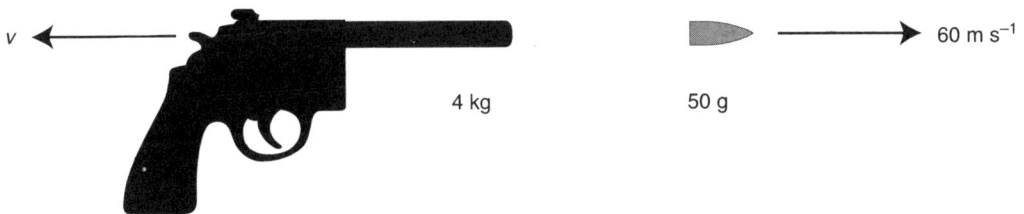

$v \leftarrow$ 4 kg   50 g $\rightarrow$ 60 m s^{-1}

Calculate the recoil speed of the gun.

Solution:

After the gun is fired the momentum of the gun (to the left) is equal to the momentum of the bullet (to the right). Momentum is a vector (since velocity is a vector) but to avoid negative numbers it is easier to consider.

$$mv \text{ (to the left)} = mv \text{ (to the right)}$$
$$4v = 0{\cdot}05 \times 60$$
$$v = \frac{0{\cdot}05 \times 60}{4}$$
$$v = \underline{\underline{0{\cdot}75 \text{ m s}^{-1}}}$$

The recoil speed of the gun is 0·75 m s^{-1} (and this is the answer).
The recoil velocity of the gun is actually −0·75 m s^{-1} since it is in the opposite direction to the bullet.

LEARN
Elastic $\Big)$ E_k is conserved collision
Explosion:
mv = mv to the left to the right

Progress to P.P. 'H' P. Page 43, nos 1– 8

IMPULSE

The impulse of a force is defined as the force multiplied by the time of contact of the force (on an object).

$$Ns \text{ ——} \quad Impulse \quad = \quad \underset{N}{F} \quad \times \quad t \quad \text{——} \quad s$$

From Newton's 2nd Law

$$F = ma$$

$$F = m\left(\frac{v - u}{t}\right)$$

$$F = \frac{m(v - u)}{t}$$

$$\underset{impulse}{F \times t} = \underset{\substack{final \\ momentum}}{mv} - \underset{\substack{initial \\ momentum}}{mu}$$

or   impulse = change in momentum

Notice that impulse can be measured in Ns or kg m s^{-1}.

When a football is kicked, the force builds up from 0 to a maximum and then the force falls to zero again when the ball leaves the foot.

The graph shows this happening.

The area under the graph is the impulse of the force.

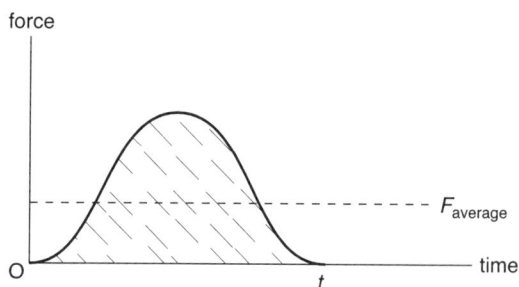

Clearly, if the time of contact increases, the average force will reduce.

Three examples of this fact are shown opposite.

Three Examples of Impulse in Action

EXAMPLE 1: *The boxer*

A punch on the chin from a bare knuckle fighter has a very short time of contact and therefore a large force.

Using a boxing glove increases the time of contact and so reduces the force.

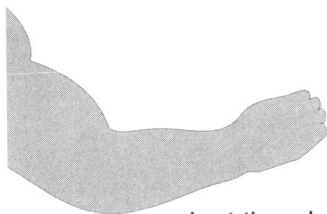

short time, large force   long time, small force

EXAMPLE 2: *The training shoe*

A hard trainer gives a short time of contact with the ground which results in a large force on the runner's heel. A soft trainer has a built-in cushion (of air or gel) designed to increase the time of contact, thereby reducing the force on the runner's heel.

short time, large force   long time, small force

EXAMPLE 3: *The pole vaulter*

A pole vaulter, after releasing the pole, can fall over 5 m.

With a cushion of sand at the bottom, there would be a short time of contact and so a large force. A modern inflatable cushion increases the time of contact and reduces the force.

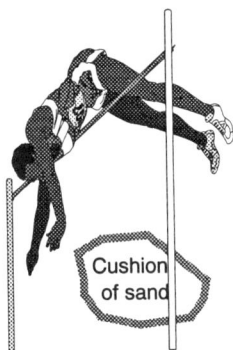

Cushion of sand

Inflatable cushion

short time, large force

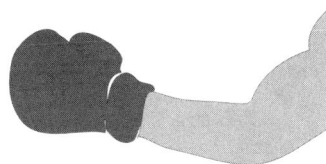

long time, small force

LEARN
$F \times t = mv - mu$

Progress to P.P. 'H' P. Page 44, nos 1–5

Apparatus

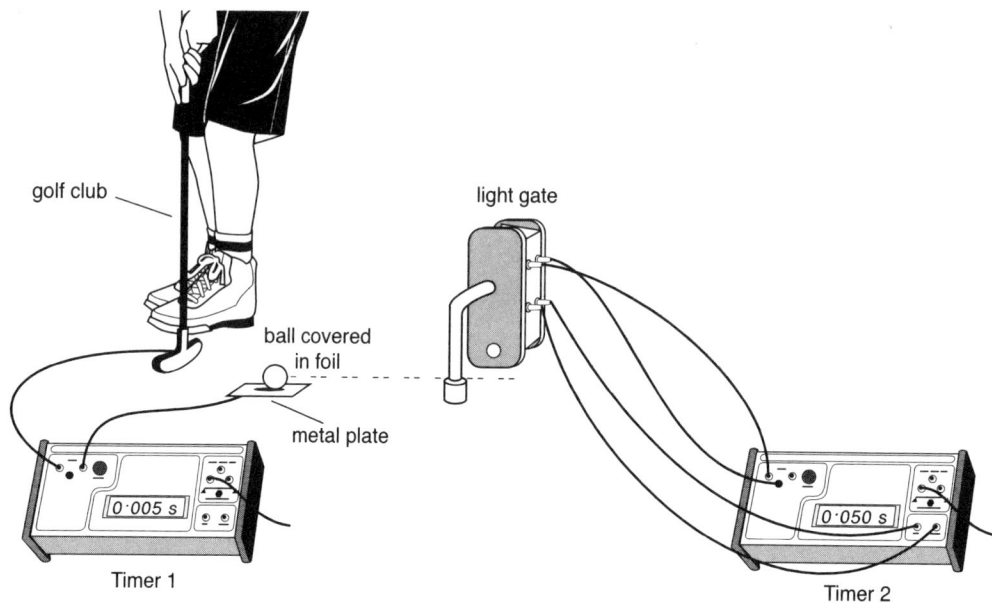

Results

Time on timer 1	=	0·005 s
Time on timer 2	=	0·050 s
Diameter of ball	=	24 mm (= 0·024 m)
Mass of ball	=	44 g (= 0·044 kg)

Experiment to find the (average) force exerted by a golf club on a golf ball

Object:

To find the (average) force exerted by a golf club on a golf ball.

Procedure:

The ball is struck horizontally so that the diameter of the ball passes through the light beam.

Aluminium foil on the ball and the metal club head touch and complete the circuit so timer 1 measures the time of contact.

Timer 2 measures the time the ball spends in the light beam. This is similar to a card on a trolley cutting a light beam, except the diameter of the ball is taking the place of the card.

Calculation:

First find velocity of ball (after being struck):

$$v \quad = \frac{\text{diameter of ball}}{\text{time on timer 2}} = \frac{0 \cdot 024}{0 \cdot 050} = 0 \cdot 48 \text{ m s}^{-1}$$

Second find force from impulse equation:

$$F \times t_1 \ = \ mv - mu$$

$$F \quad = \frac{mv}{t_1} \ (\text{since } u = 0)$$

$$F \quad = \frac{0 \cdot 044 \times 0 \cdot 48}{0 \cdot 005}$$

$$F \quad = \underline{\underline{4 \cdot 22 \text{ N}}}$$

The average force exerted on the ball is 4·22 N.

Progress to P.P. 'H' P. Page 45, nos 6–10

CHAPTER 5

DENSITY AND PRESSURE

DENSITY

A technician finds some apparatus in the store: a 1 kg mass of copper and a 1 kg mass of aluminium.

1 kg copper	1 kg aluminium

Both materials have the same mass.
Aluminium occupies a larger volume than copper.
Copper is denser than aluminium.

Definition

Density (ρ) is defined as mass per unit volume.

$$\text{kg m}^{-3} \longrightarrow \quad \rho \quad = \quad \frac{m}{V} \begin{array}{l} \text{kg} \\ \text{m}^3 \end{array}$$

Experiment to measure the density of air

In an experiment designed to measure the density of air, a container full of air is attached to a balance and the mass recorded (mass m). More air is now pumped into the container and the new mass recorded (mass M).

Subtracting the masses $(M - m)$ gives the mass of the extra air pumped into the flask.

In order to find the volume of this extra air, it is allowed to escape through a tube as shown in the diagram.

This extra air is collected in a gas jar by the downward displacement of water.

It is a simple matter to find the volume (V) of this air above the water.

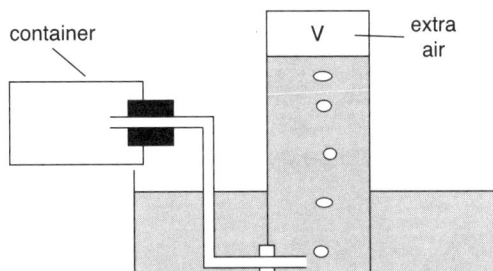

Calculation

The density of air can now be calculated from:

$$\text{density} = \frac{\text{mass of extra air}}{\text{volume}} = \frac{M - m}{V}$$

Relative densities of solid, liquid and gas

If a 1 kg lump of ice is melted, the volume of the water formed is the same as the volume of the ice. If the water is now heated and turned into steam, it will occupy a volume equal to 1000 times the volume of the water (or ice).

Hence the density of a solid = the density of a liquid = $\frac{1}{1000}$ the density of a gas

	Solid or liquid	Gas
Average spacing between molecules	x	$10x$
Relative volume	x^3	$10x \times 10x \times 10x = 1000x^3$
Relative density	y	$\dfrac{y}{1000}$

LEARN
$\rho = \dfrac{m}{V}$

PRESSURE

Pressure is defined as the force per unit area.

$$\text{Pa} \longrightarrow \boxed{p \quad = \quad \frac{F}{A}} \begin{array}{l} \text{N} \\ \\ \text{m}^2 \end{array}$$

Note: $1\ \text{Pa} = \text{N m}^{-2}$

Pressure is the force per unit area when the force is acting normal to the surface.

EXAMPLE:

Problem:

Calculate the pressure exerted by a 40 kg mass sitting on a bench. The base of the mass measures 4 m × 2 m.

Solution:

$$\begin{aligned} \text{Area} \ &= \ l \times b \\ &= \ 4 \times 2 \\ &= \ \underline{8\ \text{m}^2} \end{aligned}$$

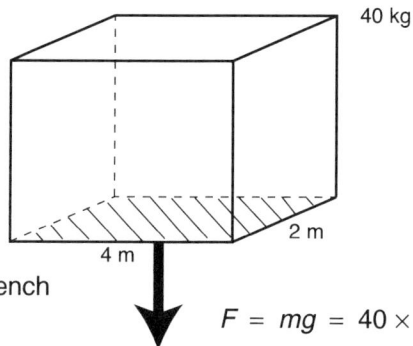

40 kg

2 m

4 m

The force acting normal to the bench
is the weight of the 40 kg mass.

$F = mg = 40 \times 9{\cdot}8 = \underline{392\ \text{N}}$

$$p \quad = \frac{F}{A}$$

$$p \quad = \frac{392}{8}$$

$$p \quad = \underline{\underline{49\ \text{N m}^{-2}}}$$

Progress to P.P. 'H' P. Page 47, nos 1–10

Pressure, depth and density

As a diver descends, the pressure on his body increases.

The pressure is due to the weight of liquid above him.

Consider the cylinder of water shown.

The pressure on the bottom area (A) due to the weight of the water above is:

$$\text{Pressure} = \frac{F}{A}$$

$$= \frac{mg}{A} \qquad [\text{Since } F = mg]$$

$$= \frac{V\rho g}{A} \qquad [\text{Since } m = V\rho]$$

$$= \frac{Ah \times \rho g}{A} \qquad [\text{Since } V = Ah]$$

$$= h\rho g$$

Pressure	$\propto$	depth
Pressure	$\propto$	density

$$\text{Pa} \longrightarrow \boxed{p = h\rho g} \longleftarrow 9\cdot8 \text{ N kg}^{-1}$$

$$\text{m} \qquad \text{kg m}^{-3}$$

In reality the pressure is ($h\rho g$ + 1 atmosphere) but the pressure on the diver before he entered the water was 1 atmosphere so only the **increase in pressure** due to the liquid has been considered. Similarly for the buoyancy calculation (below).

Buoyancy Force (Upthrust)

A cube of side 10 cm is submerged to a depth of 20 cm (see diagram). The density of water is 1000 kg m^{-3}.

The force on surface Q is greater than the force on surface P because the pressure at Q is greater than the pressure at P.

The difference in these two forces ($F_Q - F_P$) is the upthrust.

Pressure at P	Pressure at Q
$p = h\rho g$	$p = h\rho g$
$= 0\cdot1 \times 1000 \times 9\cdot8$	$= 0\cdot2 \times 1000 \times 9\cdot8$
$= 980$ Pa	$= 1960$ Pa
Force at P	**Force at Q**
$F = pA$	$F = pA$
$= 980 \times (0\cdot1 \times 0\cdot1)$	$= 1960 \times (0\cdot1 \times 0\cdot1)$
$= 9\cdot8$ N	$= 19\cdot6$ N

Upthrust is the difference between these two forces
$$\text{Upthrust} = F_Q - F_P = 19\cdot6 - 9\cdot8 = 9\cdot8 \text{ N}$$

Progress to P.P. 'H' P. Page 49, nos 11–15

LEARN
$\text{Pressure} = \dfrac{\text{force}}{\text{area}}$
$p = h\rho g$

CHAPTER 6

GAS LAWS

Apparatus

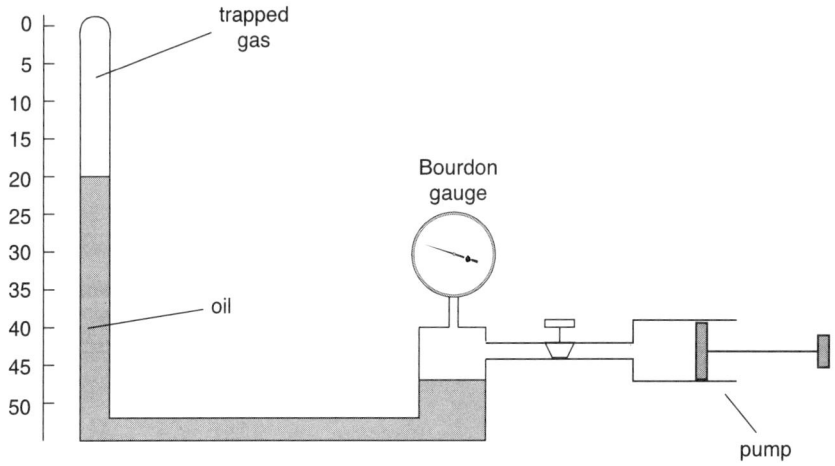

Graphs

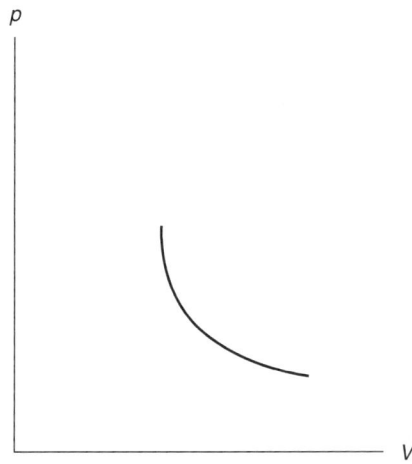

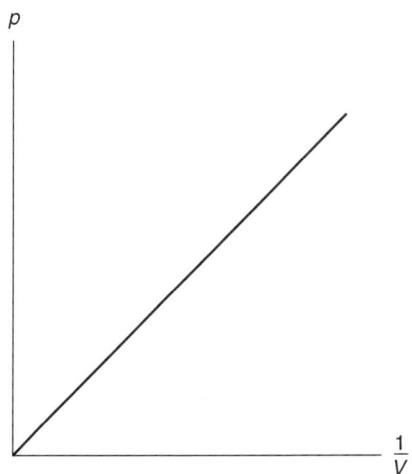

BOYLE'S LAW

Object

To find a relationship between pressure and volume for a fixed mass of gas at constant temperature.

Procedure

In the apparatus opposite, the pump is used to increase the pressure (measured on the Bourdon gauge). The oil moves up the tube, compressing the gas into a smaller volume (measured on the scale).

Results

$\dfrac{1}{V}$ / cm^{-3}	$p \times 10^5$ / Pa	V / cm^3	$p \times V$ / 10^5
0·025	0·99	40·25	39·8
0·025	1·00	40·00	40·0
0·027	1·09	37·00	40·3
0·029	1·12	35·00	39·2
0·035	1·40	28·60	40·0
0·037	1·50	27·00	40·0

Conclusion

From the second graph (and from the final column above)

$$p \propto \frac{1}{V}$$

$$p = k\frac{1}{V} \qquad (k = \text{a constant})$$

$$\boxed{\therefore\ pV = k}$$ Provided T constant, fixed mass

LEARN
$p \propto \dfrac{1}{V}$

Progress to P.P. 'H' P. Page 51, nos 1–9

Apparatus

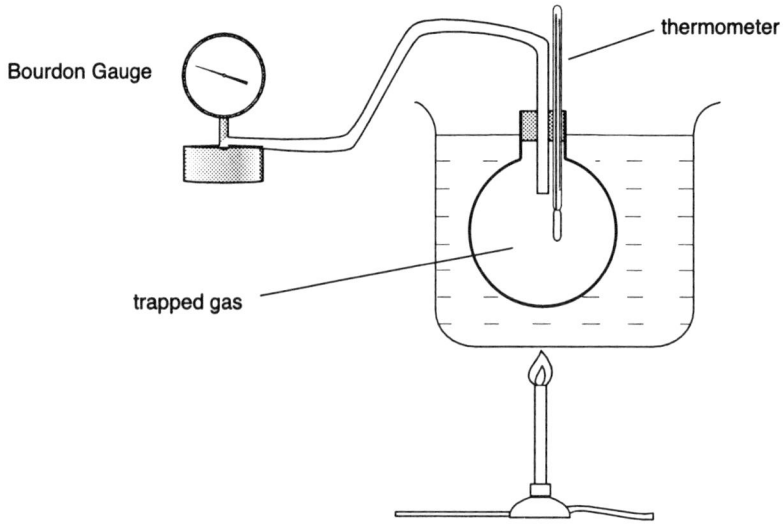

Graphs

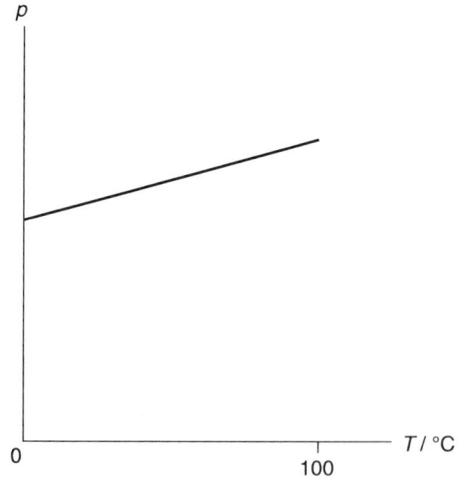

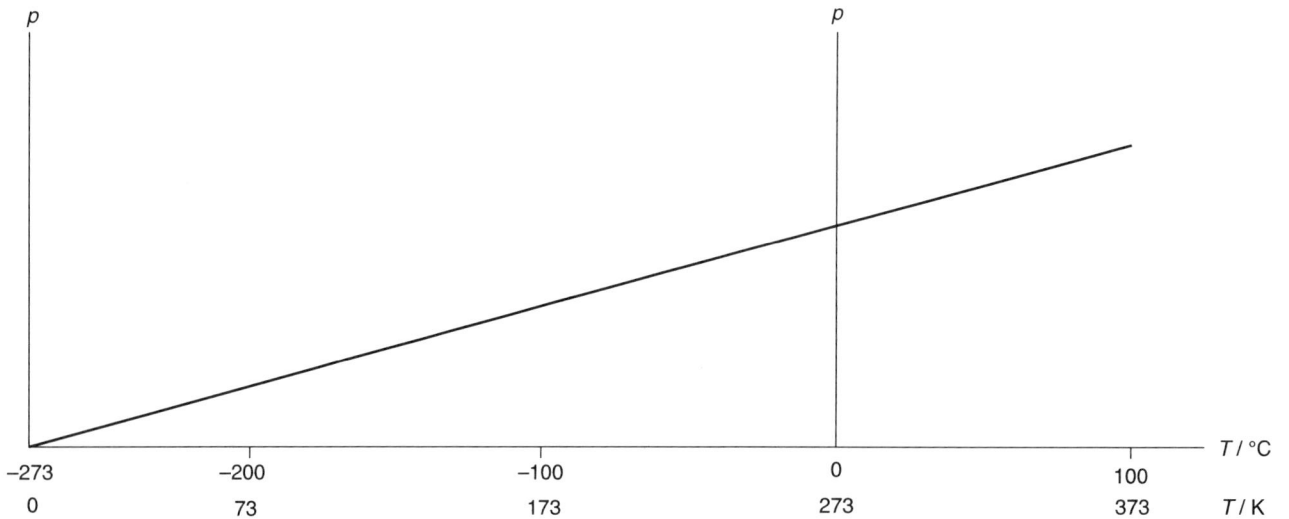

PRESSURE LAW

Object

To find a relationship between pressure and temperature for a fixed mass of gas at constant volume.

Procedure

In the apparatus opposite, the water is heated and the pressure increases. Pressure readings (on the Bourdon gauge) are recorded at different temperatures (as measured by the thermometer).

Results

$T\,/\,°C$	$p \times 10^5\,/\,Pa$
16	0·99
17	1·00
55	1·10
71	1·13
76	1·14
89	1·18

Conclusion

The graph of p against T (°C) is a straight line but it does not go through the origin. When this line is extrapolated back to cut the x-axis, this is the zero of temperature called the Absolute Zero and $p \propto T$ on this new scale (now called the Kelvin temperature scale).

$$p \quad \propto T_K$$
$$p \quad = kT_K \qquad (k = \text{a constant})$$

$$\boxed{\frac{p}{T_K} = k}$$ Provided V constant, fixed mass

LEARN
$p \propto T_K$

Progress to P.P. 'H' P. Page 53, nos 1–4

Apparatus

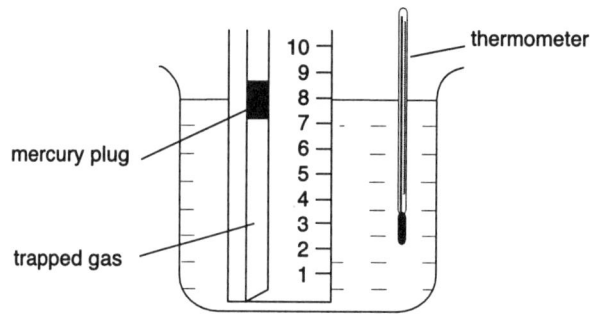

Graphs

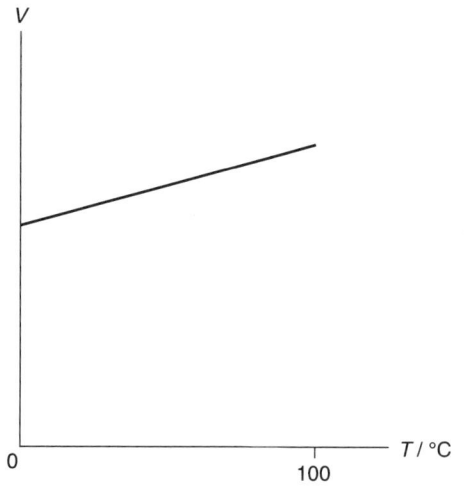

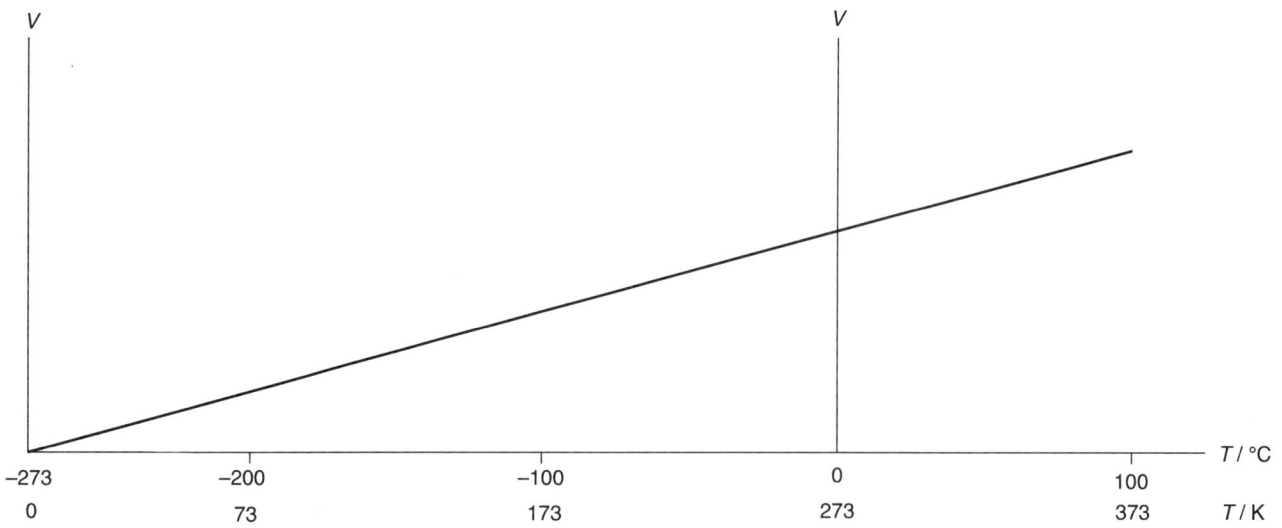

CHARLES' LAW

Object

To find a relationship between volume and temperature for a fixed mass of gas at constant pressure.

Procedure

In the apparatus opposite, the temperature of the water is altered (measured on the thermometer). When heated, the gas expands and pushes the mercury plug up the tube. The linear scale gives a measure of the volume of the trapped gas (since the cross-sectional area of the tube is constant).

Results

T / °C	V / cm
1	5·8
27	6·0
64	6·5
77	6·8
100	7·0

Conclusion

The graph of V against T (°C) is a straight line but it does not go through the origin. When this line is extrapolated back to cut the x-axis, this is the zero of temperature called the Absolute Zero and $V \propto T$ on this new scale (now called the Kelvin temperature scale).

$$V \propto T_K$$
$$V = kT_K \qquad (k = \text{a constant})$$

$$\boxed{\frac{V}{T_K} = k} \qquad \text{Provided } p \text{ constant, fixed mass}$$

LEARN
$V \propto T_K$

Progress to P.P. 'H' P. Page 54, nos 1–4

THE GENERAL GAS EQUATION

From Boyle's Law (*p* and *V* with *T* constant)

$$p \propto \frac{1}{V}$$

$\underline{\underline{p_1 V_1 = p_2 V_2}}$ for a fixed mass of gas *T* constant.

From Pressure Law (*p* and *T* with *V* constant)

$$p \propto T$$

$\underline{\underline{\dfrac{p_1}{T_1} = \dfrac{p_2}{T_2}}}$ for a fixed mass of gas *V* constant.

From Charles' Law (*V* and *T* with *p* constant)

$$V \propto T$$

$\underline{\underline{\dfrac{V_1}{T_1} = \dfrac{V_2}{T_2}}}$ for a fixed mass of gas *p* constant.

In each case, one of the three variables *p*, *V* and *T* was kept constant. What if none of the three is kept constant?

E.g., "what happens to the pressure of a fixed mass of gas if the volume is decreased and the temperature is increased at the same time?"

Combining the three laws:

$$p \propto \frac{T}{V}$$

$$p = \frac{kT}{V} \quad \text{(where } k \text{ is a constant)}$$

$$\frac{pV}{T} = k$$

The General Gas Equation

$$\boxed{\frac{p_1 V_1}{T_1} = \frac{p_2 V_2}{T_2}} \quad \text{for a fixed mass of gas}$$

p_1 = initial pressure (Pa) p_2 = final pressure (Pa)

V_1 = initial volume (m^3) V_2 = final volume (m^3)

T_1 = initial temperature (K) T_2 = final temperature (K)

The three laws, (Boyle's, Pressure, Charles'), are just special cases of the general gas equation.

Worked Example on the General Gas Equation

Problem

A fixed mass of gas in a sealed chamber is heated from 67 °C to 107 °C while at the same time the gas is compressed from 34 cm^3 to 28 cm^3. If the gas was originally at a pressure of 1.0×10^5 Pa, calculate the pressure of the gas after heating and compression.

Solution

First convert the Centigrade temperatures into Kelvin temperatures.

$$T_1 = 67 \text{ °C} = (67 + 273)\text{K} = 340 \text{ K}$$

$$T_2 = 107 \text{ °C} = (107 + 273)\text{K} = 380 \text{ K}$$

$$p_1 = 1.0 \times 10^5 \text{ Pa} \qquad\qquad p_2 = ?$$

$$V_1 = 34 \text{ cm}^3 \qquad\qquad V_2 = 28 \text{ cm}^3$$

$$T_1 = 340 \text{ K} \qquad\qquad T_2 = 380 \text{ K}$$

Applying the general gas equation:

$$\frac{p_1 V_1}{T_1} = \frac{p_2 V_2}{T_2}$$

$$\frac{1.0 \times 10^5 \times 34}{340} = \frac{p_2 \times 28}{380}$$

$$p_2 = \frac{1.0 \times 10^5 \times 34 \times 380 \text{ Pa}}{340 \times 28}$$

$$p_2 = \frac{38}{28} \times 10^5 \text{ Pa}$$

$$p_2 = \underline{\underline{1.36 \times 10^5 \text{ Pa}}}$$

i.e., the pressure has increased by about a third.

Progress to P.P. 'H' P. Page 55, nos 1–5

LEARN
$\dfrac{p_1 V_1}{T_1} = \dfrac{p_2 V_2}{T_2}$

Molecules move at the same velocity

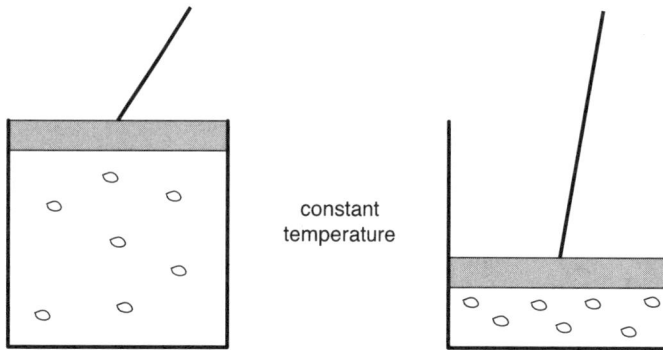

constant
temperature

molecules
slow

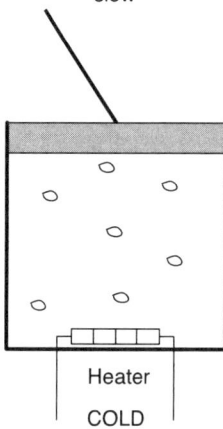

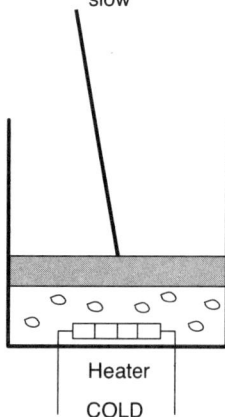

molecules
fast

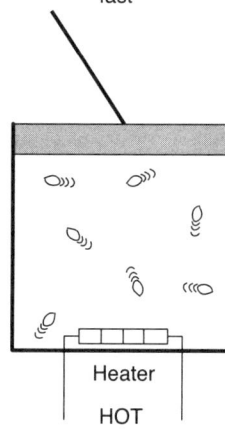

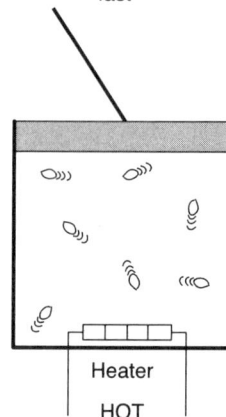

constant
volume

Heater

COLD

Heater

HOT

molecules
slow

molecules
fast

constant
pressure

Heater

COLD

Heater

HOT

KINETIC THEORY

The gas laws can be explained by considering molecular motion.

Boyle's Law (p and V with T constant)

When a fixed mass of gas (i.e., a fixed number of molecules) is compressed in a sealed container,

> the volume decreases
>
> the molecules are moving with the same velocity
>
> they make more collisions per second
>
> the pressure increases.

Pressure Law (p and T with V constant)

When a fixed mass of gas (i.e., a fixed number of molecules) is heated in a sealed container,

> the temperature increases
>
> the average E_k of the molecules increases
>
> the molecules move faster
>
> they hit the walls harder and more often
>
> the pressure increases.

Charles' Law (V and T with p constant)

When a fixed mass of gas (i.e., a fixed number of molecules) is heated in a sealed container,

> the temperature increases
>
> the average E_k of the molecules increases
>
> the molecules move faster
>
> the gas must expand to maintain pressure
>
> the volume increases.

LEARN
When molecules are heated they move faster.

UNIT 2

ELECTRICITY AND ELECTRONICS

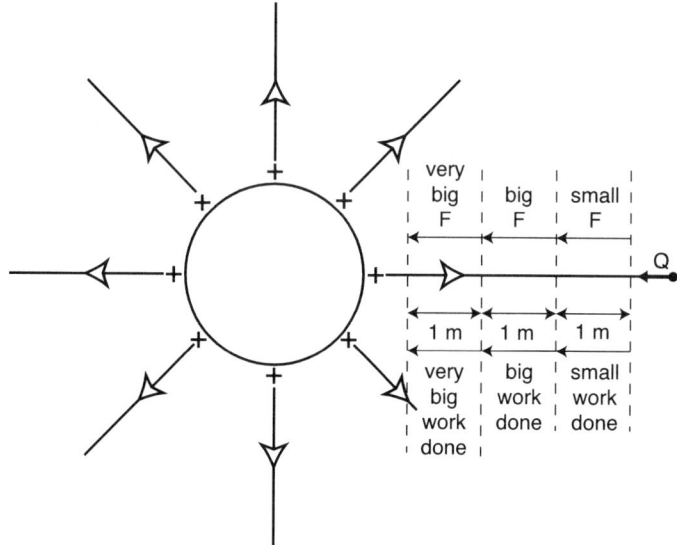

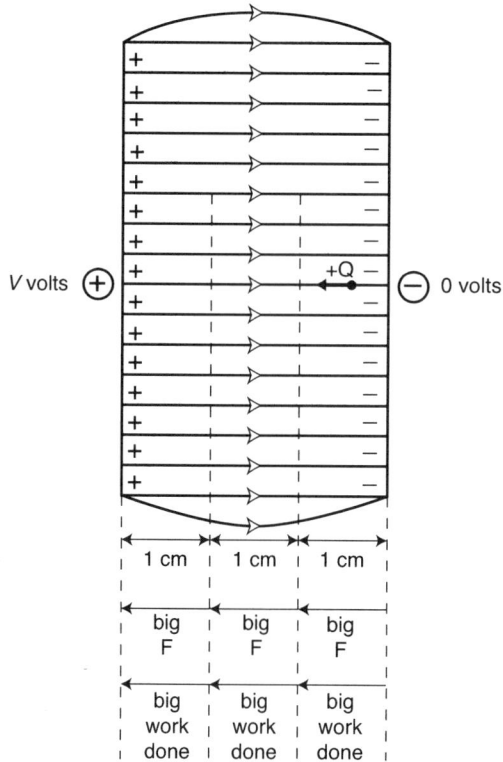

CHAPTER 7

ELECTRIC FIELDS AND RESISTORS IN CIRCUITS

CHARGES MOVING IN ELECTRIC FIELDS

Radial Field

The diagram opposite shows the field round the dome of a Van de Graaf generator. The electric field is strongest at the surface of the dome and it becomes weaker as distance from the dome increases. There is no field inside the dome.

A positive charge Q is forced towards the dome. Work is done (force × distance) but the field becomes stronger so more work per meter must be done as the charge approaches the dome.

Uniform Field

The diagram opposite shows the field between two parallel plates.

The electric field has the same strength everywhere between the plates.

A positive charge Q is forced towards the positive plate. The work done to travel the first centimeter is the same as the work done to travel the second centimeter, which is the same as the work done to travel the third centimeter. This is because the electric field strength is constant.

The work done (W) in moving charge Q through V volts is:

$$W = QV$$

Rearranging this equation gives $V = \dfrac{W}{Q}$ which leads to the definition of the volt, i.e., $1\ V = 1\ \text{J C}^{-1}$.

LEARN
$W = QV$

Progress to P.P. 'H' P. Page 56, nos 1– 3

CHARGES MOVING IN ELECTRIC FIELDS

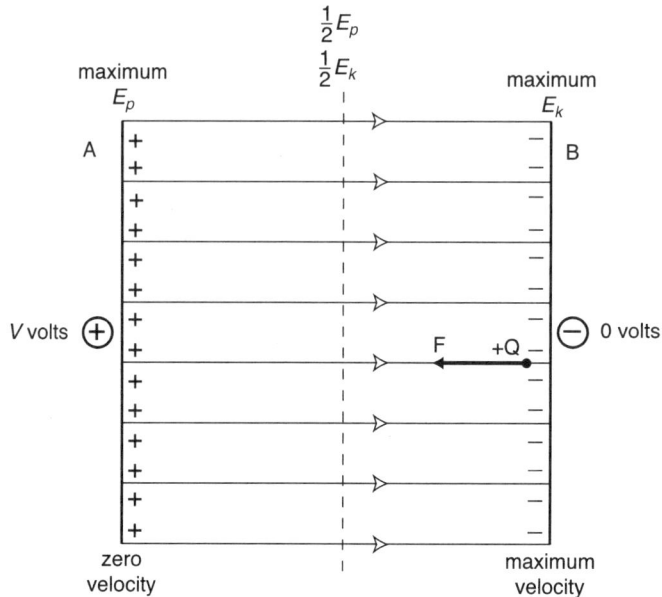

In moving a charge $+Q$ from B to A (above), work must be done to overcome the repulsion.

The work done to move charge $+Q$ from B to A is $W = QV$.

Therefore charge $+Q$ now has potential energy equal to that work done.

$$\therefore \text{ potential energy at A } = QV$$

If charge $+Q$ now moves from A to B, this potential energy is converted into kinetic energy. Half way between the plates, half the energy is potential and half is kinetic. When the charge strikes plate B, all the energy is kinetic and the velocity is a maximum.

$$\therefore \text{ kinetic energy at B } = \frac{1}{2} mv^2$$

When the charge strikes B it has kinetic energy equal to the potential energy it had at A.

$$\text{potential energy at A } = \text{ kinetic energy at B}$$

$$\boxed{QV = \frac{1}{2} mv^2}$$

WORKED EXAMPLE

Problem

An electron is accelerated (from rest) in an electron gun by a voltage of 2500 V.

(a) Calculate the potential energy of the electron at X.

(b) What is the kinetic energy of the electron at Y?

(c) Calculate the maximum velocity of the electron.

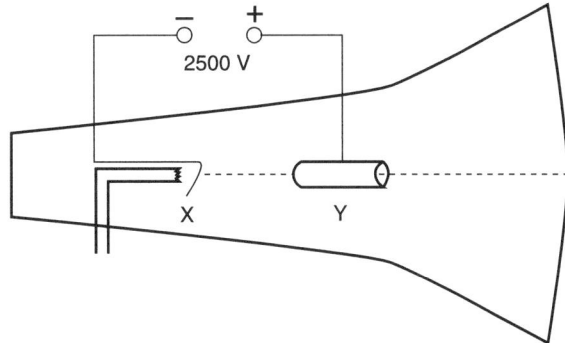

Solution

(a) E_p = QV

   = $1 \cdot 6 \times 10^{-19} \times 2500$

   = $4 \cdot 0 \times 10^{-16}$ J

(b) Same as the potential energy at X

   = $4 \cdot 0 \times 10^{-16}$ J

(c) $\frac{1}{2} mv^2$ = QV

   v^2 = $\frac{2QV}{m}$

   v^2 = $\dfrac{2 \times 1 \cdot 6 \times 10^{-19} \times 2500}{9 \cdot 11 \times 10^{-31}}$

   v^2 = $8 \cdot 78 \times 10^{14}$

   v = $2 \cdot 96 \times 10^7$ m s^{-1}

LEARN
$QV = \dfrac{1}{2} mv^2$

Progress to P.P. 'H' P. Page 57, nos 4– 12

E.M.F. AND INTERNAL RESISTANCE

The e.m.f. (electromotive force) of a cell is the voltage between the terminals on open circuit.

Every cell has internal resistance so when a current passes through the internal resistance some voltage is dropped across it.

Consequently the voltage between the terminals (A and B) of the cell ($V_{\text{t.p.d.}}$) will always be less than the e.m.f.

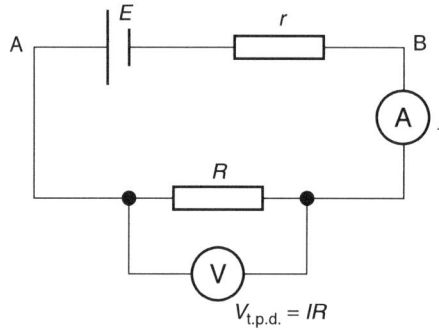

Equation

In the circuit above, the total resistance is $R_T = R + r$.

Applying Ohm's Law to the whole circuit:

$$E = IR_T$$
$$E = I(R + r)$$
$$E = IR + Ir$$
$$E = V_{\text{t.p.d.}} + Ir$$

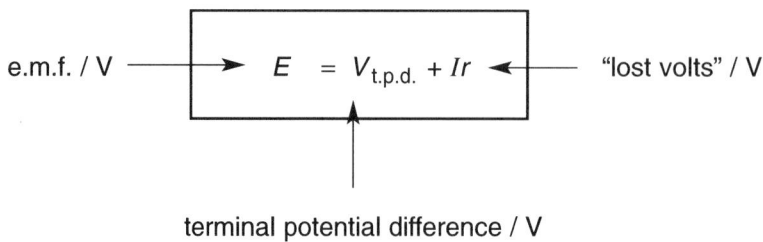

e.m.f. / V ———▶ $E = V_{\text{t.p.d.}} + Ir$ ◀——— "lost volts" / V

terminal potential difference / V

INTERNAL RESISTANCE WORKED EXAMPLE

Problem

In the circuit shown find:

(a)   the reading on the voltmeter;

(b)   the "lost volts";

(c)   the internal resistance, r.

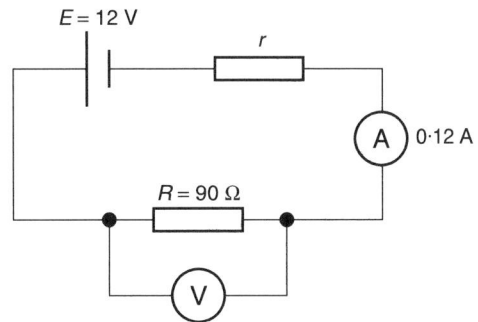

Solution

(a) $V = IR$
      $V = 0.12 \times 90$
      $V = \underline{\underline{10.8 \text{ V}}}$

(b)   Lost volts $= 12 - 10.8$
                  $= \underline{\underline{1.2 \text{ V}}}$

(c) $r = \dfrac{\text{lost volts}}{I}$

      $r = \dfrac{1.2}{0.12}$

      $r = \underline{\underline{10 \, \Omega}}$

LEARN
$E = V_{\text{t.p.d.}} + Ir$

Progress to P.P. 'H' P. Page 60, nos 1–11

Apparatus

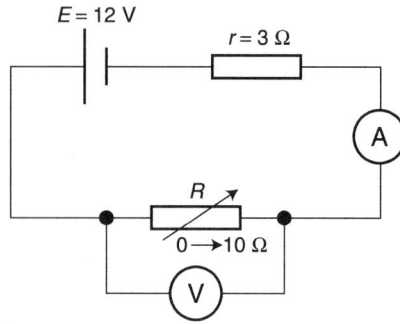

Graph

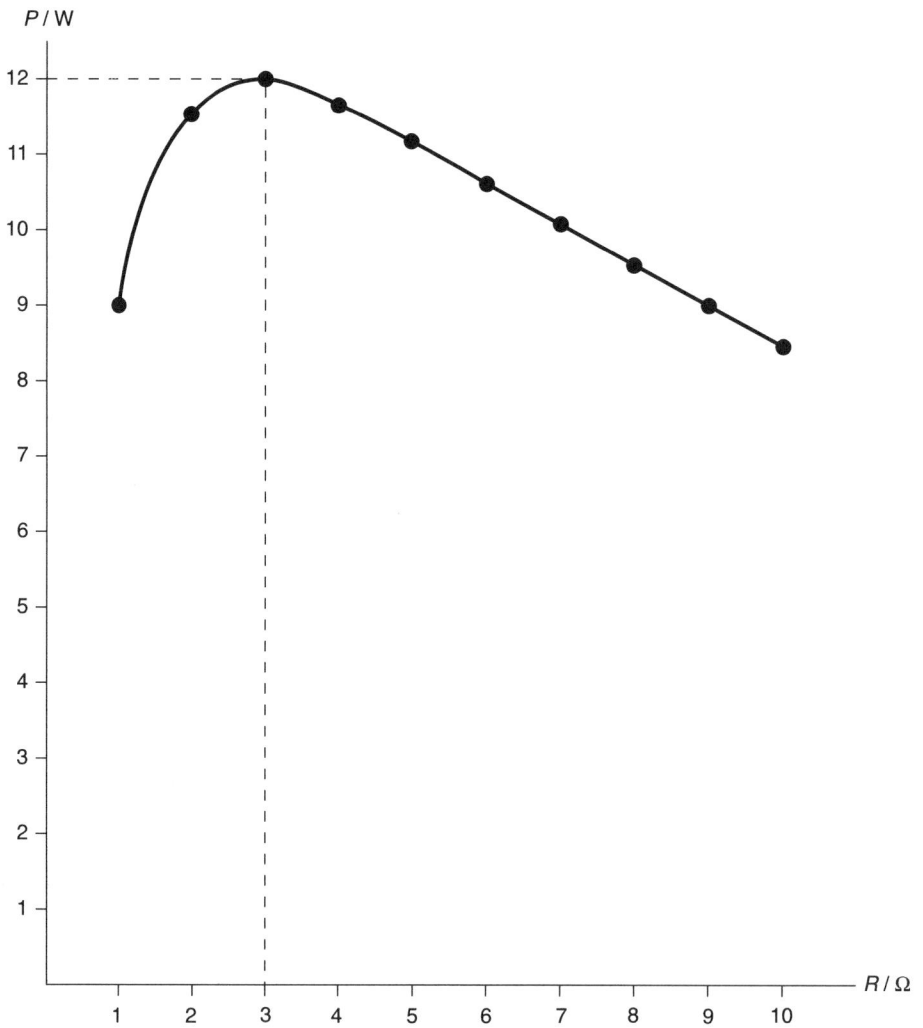

Maximum power transfer

Referring to the circuit opposite:

To obtain a high voltage ($V_{\text{t.p.d.}}$) <u>R is increased</u> but the current gets small.

To obtain a high current <u>R is reduced</u> but the voltage ($V_{\text{t.p.d.}}$) gets small.

It is not possible to obtain a large voltage and a large current at the same time. The compromise is a "medium" voltage and a "medium" current to deliver maximum power to the external resistor R.

Method

By changing R from 1 to 10 Ω and taking corresponding readings on the voltmeter and ammeter, the following table of results can be produced.

Results

R / Ω	r / Ω	R_T / Ω		I / A	$V_{\text{t.p.d.}} / V$	$P = IV / W$
1	3	4		3·00	3·00	9·00
2	3	5		2·40	4·80	11·52
3	3	6		2·00	6·00	12·00
4	3	7		1·71	6·84	11·70
5	3	8		1·50	7·50	11·25
6	3	9		1·33	7·98	10·61
7	3	10		1·20	8·40	10·08
8	3	11		1·09	8·72	9·50
9	3	12		1·00	9·00	9·00
10	3	13		0·92	9·20	8·46

Conclusion

From the graph opposite (or from the table of results), maximum power is transferred to the load resistor R when R = 3 Ω.

I.e., maximum power is transferred to the external (load) resistor R when <u><u>R = r.</u></u>

LEARN
Condition for maximum power transfer
R = r

Progress to P.P. 'H' P. Page 66, no 12

RESISTORS IN CIRCUITS

Proof of Resistors in Series Equation

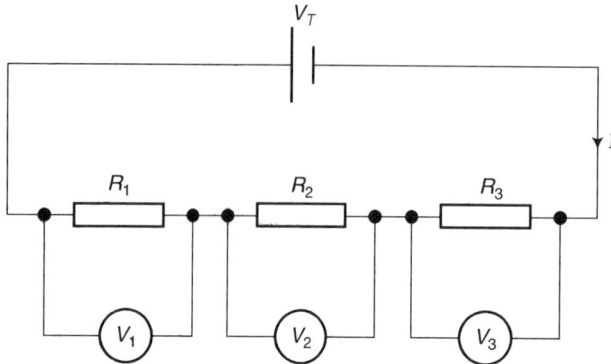

The current through each resistor is the same (I). By conservation of energy:

$$V_T = V_1 + V_2 + V_3$$
$$IR_T = IR_1 + IR_2 + IR_3$$
$$I\!\!\!/\, R_T = I\!\!\!/\, (R_1 + R_2 + R_3)$$
$$R_T = R_1 + R_2 + R_3$$

Proof of Resistors in Parallel Equation

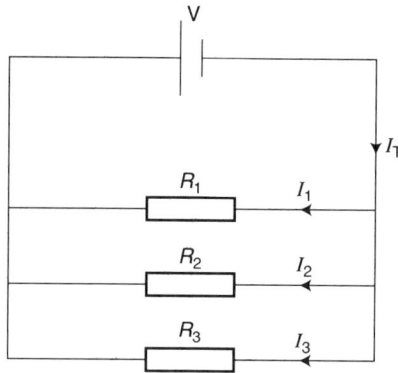

The voltage across each resistor is the same (V). By conservation of charge:

$$I_T = I_1 + I_2 + I_3$$
$$\frac{V}{R_T} = \frac{V}{R_1} + \frac{V}{R_2} + \frac{V}{R_3}$$
$$\frac{V\!\!\!/}{R_T} = V\!\!\!/\left(\frac{1}{R_1} + \frac{1}{R_2} + \frac{1}{R_3}\right)$$
$$\frac{1}{R_T} = \frac{1}{R_1} + \frac{1}{R_2} + \frac{1}{R_3}$$

Worked Example 1

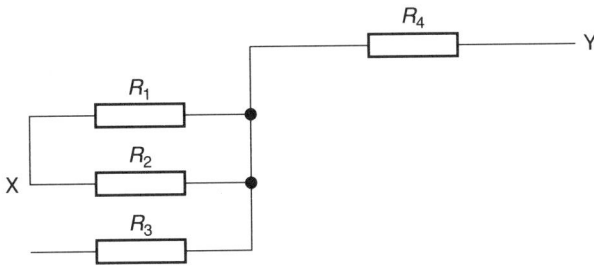

Problem

All resistors are 100 Ω. Find the resistance between X and Y.

Solution

This needs to be solved in two steps.

Step ① (parallel section)

$$\frac{1}{R_T} = \frac{1}{R_1} + \frac{1}{R_2}$$

$$\frac{1}{R_T} = \frac{1}{100} + \frac{1}{100}$$

$$\frac{1}{R_T} = \frac{2}{100}$$

$$R_T = 50 \ \Omega$$

Step ②

$$R_{Total} = R_{parallel\ section} + R_4$$
$$R_T = 50 + 100$$
$$R_T = 150 \ \Omega$$

(Note: R_3 is not in the circuit)

Worked Example 2

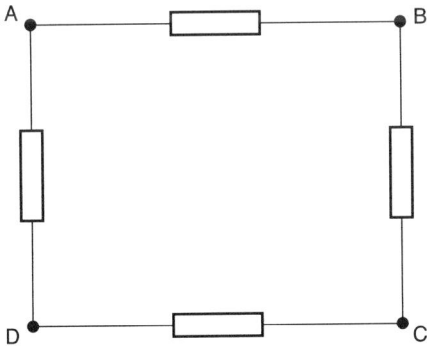

This is the same as this

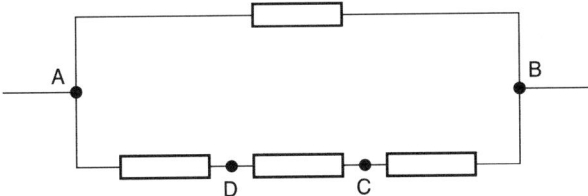

Problem

All resistors are 100 Ω. Find the resistance between A and B.

Solution

$$\frac{1}{R_T} = \frac{1}{R_1} + \frac{1}{R_2}$$

$$\frac{1}{R_T} = \frac{1}{100} + \frac{1}{300}$$

$$\frac{1}{R_T} = \frac{3 + 1}{300}$$

$$\frac{1}{R_T} = \frac{4}{300}$$

$$R_T = 75 \ \Omega$$

LEARN
Series
$R_T = R_1 + R_2 + R_3$
Parallel
$\frac{1}{R_T} = \frac{1}{R_1} + \frac{1}{R_2} + \frac{1}{R_3}$

Progress to P.P. 'H' P. Page 67, nos 1–8

Worked Example 3

Problem

In the circuit shown:

(a) calculate the reading on the ammeter when S is open;

(b) calculate the reading on the ammeter when S is closed.

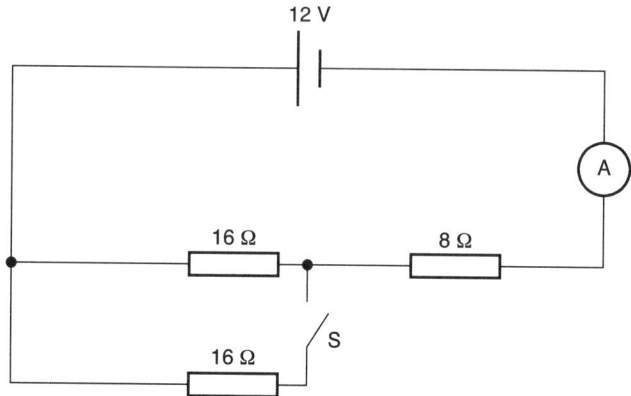

Solution

(a) Step ①: find total resistance.

$$R_T = R_1 + R_2$$
$$R_T = 16 + 8$$
$$\underline{R_T = 24 \; \Omega}$$

Step ②: find current on ammeter.

$$I = \frac{V}{R}$$
$$I = \frac{12}{24}$$
$$\underline{\underline{I = 0{\cdot}5 \; A}}$$

(b) Step ①: find resistance of parallel section.

$$\frac{1}{R_T} = \frac{1}{R_1} + \frac{1}{R_2}$$
$$\frac{1}{R_T} = \frac{1}{16} + \frac{1}{16}$$
$$\frac{1}{R_T} = \frac{2}{16}$$
$$\underline{R_T = 8 \; \Omega}$$

Step ②: find resistance of whole circuit.

$$R_T = R_{\text{parallel section}} + R$$
$$R_T = 8 + 8$$
$$\underline{R_T = 16 \; \Omega}$$

Step ③: find current on ammeter.

$$I = \frac{V}{R}$$
$$I = \frac{12}{16}$$
$$\underline{\underline{I = 0{\cdot}75 \; A}}$$

Progress to P.P. 'H' P. Page 69, nos 9–14

OHMS LAW AND POWER (ADVANCED PROBLEMS)

Some revision of Ohm's Law and Power at 'S' Grade level is advised before the student progresses to these advanced problems.

Worked Example

Problem

(a) Find R.

(b) Find the reading on each voltmeter.

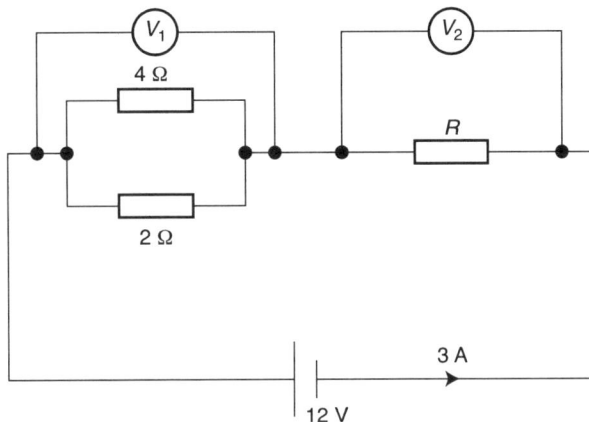

Solution

(a) Step ①:
     total resistance.

$$R_T = \frac{V}{I}$$
$$R_T = \frac{12}{3}$$
$$\underline{\underline{R_T = 4\ \Omega}}$$

Step ②:
resistance of parallel section.

$$\frac{1}{R_P} = \frac{1}{R_1} + \frac{1}{R_2}$$
$$\frac{1}{R_P} = \frac{1}{4} + \frac{1}{2}$$
$$\frac{1}{R_P} = \frac{3}{4}$$
$$\underline{\underline{R_P = \frac{4}{3} = 1\cdot33\ \Omega}}$$

Step ③:

$$R = \text{total } R - \text{parallel } R$$
$$R = R_T - R_P$$
$$R = 4 - 1\cdot33$$
$$\underline{\underline{R = 2\cdot67\ \Omega}}$$

(b) Step ①: voltage on V_2.

$$V = IR$$
$$V = 3 \times 2\cdot67$$
$$\underline{\underline{V_2 = 8\cdot01\ \text{V}}}$$

Step ②: V_1.

$$V_1 = V_{supply} - V_2$$
$$V_1 = 12 - 8\cdot01$$
$$\underline{\underline{V_1 = 3\cdot99\ \text{V}}}$$

Alternatively $V_1 = IR_P$

$$V_1 = 3 \times 1\cdot33$$
$$\underline{\underline{V_1 = 3\cdot99\ \text{V}}}$$

LEARN
$V = IR$
$P = VI$

Progress to P.P. 'H' P. Page 71, nos 1–10

THE WHEATSTONE AND METER BRIDGES

In an Ohm's Law circuit used to find an unknown resistance there is a built-in error. The voltmeter and the ammeter cannot both be correct at the same time.

Wheatstone developed his bridge circuit so that it did not need a voltmeter or ammeter to measure anything. He used four resistors and provided he knew three of them he could calculate the fourth (unknown resistor).

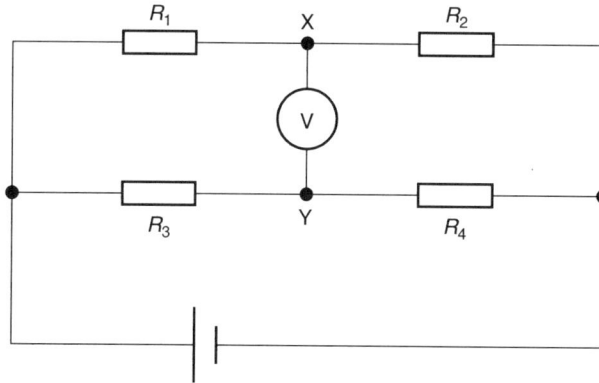

The circuit above is balanced when the meter reads zero, i.e., the potential of point X is equal to the potential of point Y.

When balanced:

$$\frac{R_1}{R_2} = \frac{R_3}{R_4}$$

Worked Example

Problem

The circuit shown is balanced.

(a) What is the reading on the voltmeter?

(b) Given the figures:

R_1 = 200 Ω

R_2 = 800 Ω

R_3 = 1000 Ω

calculate R_4.

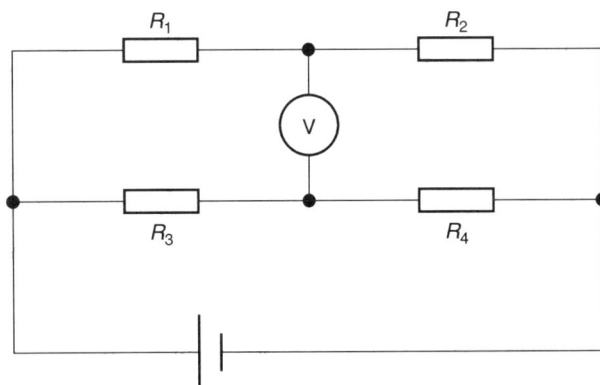

Solution

(a) 0 V

(b)
$$\frac{R_1}{R_2} = \frac{R_3}{R_4}$$

$$\frac{200}{800} = \frac{1000}{R_4}$$

$$200R_4 = 800 \times 1000$$

$$R_4 = \frac{800 \times 1000}{200}$$

$$\underline{\underline{R_4 = 4000 \ \Omega}}$$

LEARN
$\dfrac{R_1}{R_2} = \dfrac{R_3}{R_4}$

Progress to P.P. 'H' P. Page 74, nos 1–6

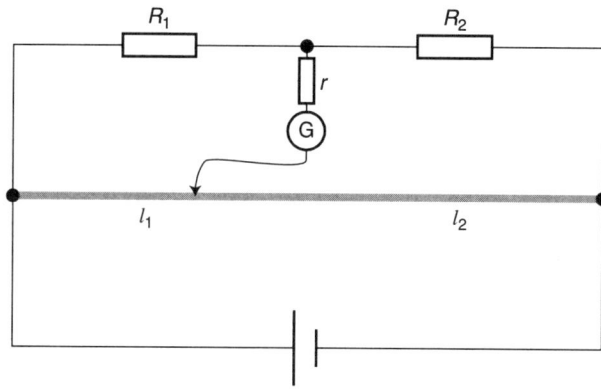

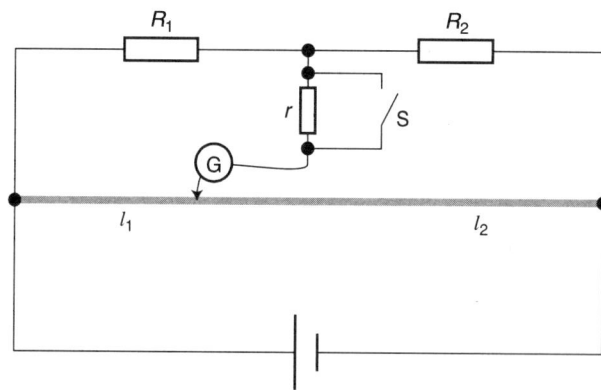

THE METER BRIDGE

The meter bridge is exactly the same as a Wheatstone bridge where two of the resistors have been replaced with a 1 m length of resistance wire (see circuit opposite).

R_3 and R_4 in the Wheatstone circuit are now two lengths of wire (which add up to 1 m).

Equation

From the Wheatstone bridge $\dfrac{R_1}{R_2} = \dfrac{R_3}{R_4}$ (when balanced)

Replacing R_3 and R_4 with l_1 and l_2 gives

$$\boxed{\dfrac{R_1}{R_2} = \dfrac{l_1}{l_2}}$$

Protecting the Meter

A galvanometer is a very sensitive ammeter and a large surge of current will seriously damage it.

To protect the galvanometer, a resistor (r) is connected in series with it.

However, it is advisable to remove this from the circuit to achieve perfect balance.

With S open, the slider is moved until the circuit (opposite) is balanced. Then the switch is closed to bypass the resistor, r.

Finally any minor adjustments are made for perfect balance (galvanometer reads 0).

LEARN
$\dfrac{R_1}{R_2} = \dfrac{l_1}{l_2}$

Worked Example

Problem

The meter bridge shown is balanced when $l_1 = 80$ cm.

(a) What is the reading on the voltmeter?

(b) Given that $R_1 = 2$ kΩ, calculate R_2.

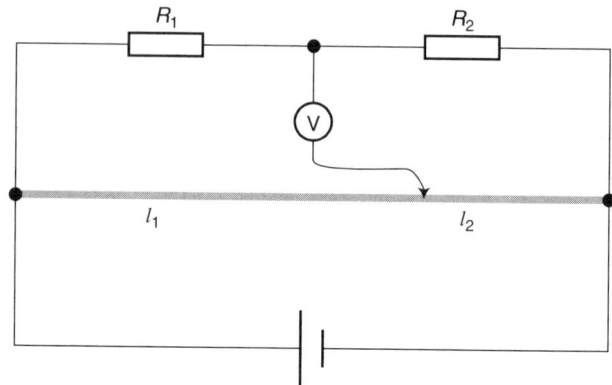

Solution

(a)   0 V

(b)   If $l_1 = 80$ cm then l_2 must be 20 cm.

$$\frac{R_1}{R_2} = \frac{l_1}{l_2}$$

$$\frac{2000}{R_2} = \frac{80}{20}$$

$$\therefore 80R_2 = 20 \times 2000$$

$$R_2 = \frac{20 \times 2000}{80}$$

$$\underline{\underline{R_2 = 500 \ \Omega}}$$

Progress to P.P. 'H' P. Page 76, nos 7–12

THE UNBALANCED WHEATSTONE BRIDGE

A Wheatstone Bridge circuit is set up as shown in the diagram and R_4 is adjusted until the bridge is balanced. At balance $R_4 = 1000\ \Omega$.

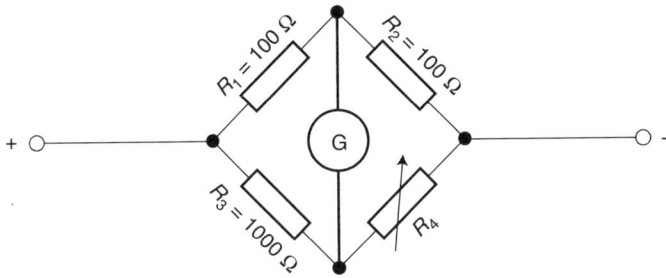

R_4 is now increased and decreased by constant amounts and the reading on the galvanometer is noted.

A graph of galvanometer current against change in R_4 produces a straight line graph through the origin.

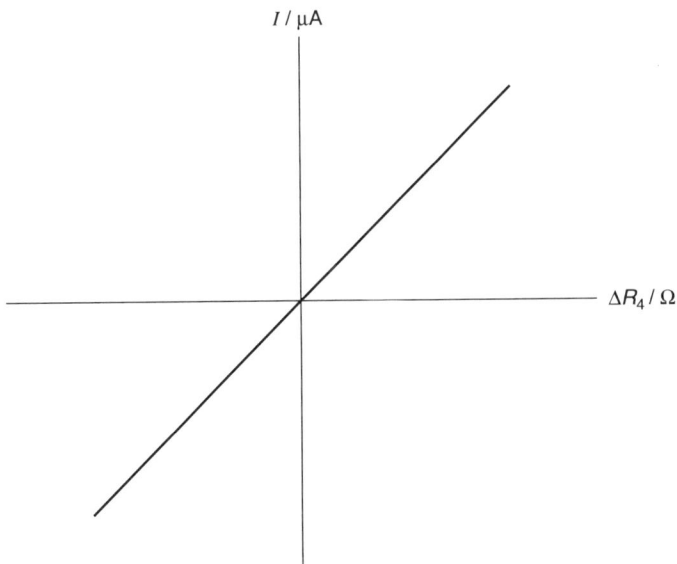

Conclusion $\boxed{I \propto \Delta R_4}$

Current is proportional to change in R_4. However, this only applies for small changes in R_4 (the graph is only a straight line near the origin).

LEARN
$I \propto \Delta R_4$ Or using a voltmeter: $V \propto \Delta R_4$

Progress to P.P. 'H' P. Page 78, nos 1–5

CHAPTER 8

ALTERNATING CURRENT AND VOLTAGE

OSCILLOSCOPES AND VOLTAGE

The centre line of the oscilloscope is 0 V and all the voltages are measured with reference to this line.

Measuring d.c.

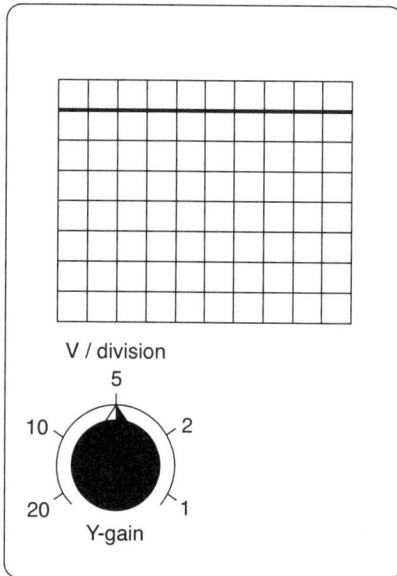

A d.c. voltage is constant. This voltage is three divisions above the centre.

$$\text{Voltage} = \text{number of divisions} \times \text{voltage per division}$$
$$= 3 \times 5$$
$$= \underline{\underline{15\ V}}$$

Measuring a.c.

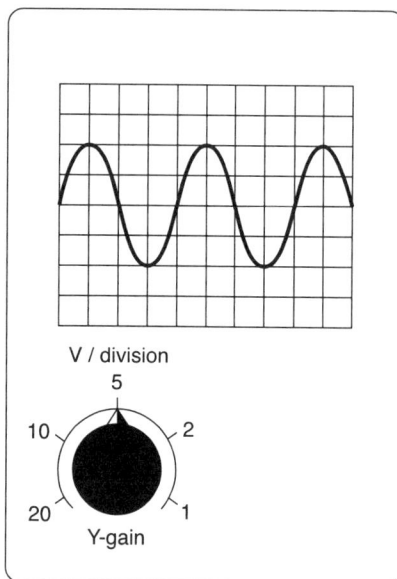

An a.c. voltage varies. The only fixed point is the peak voltage. This peak voltage is two divisions above the centre.

$$\text{Peak voltage} = \text{number of divisions} \times \text{voltage per division}$$
$$= 2 \times 5$$
$$= \underline{\underline{10\ V}}$$

Progress to P.P. 'H' P. Page 80, nos 1–5

OSCILLOSCOPES AND FREQUENCY

The oscilloscope can be used to measure the frequency of an a.c. supply.

This depends on first finding the time for 1 wave (period).

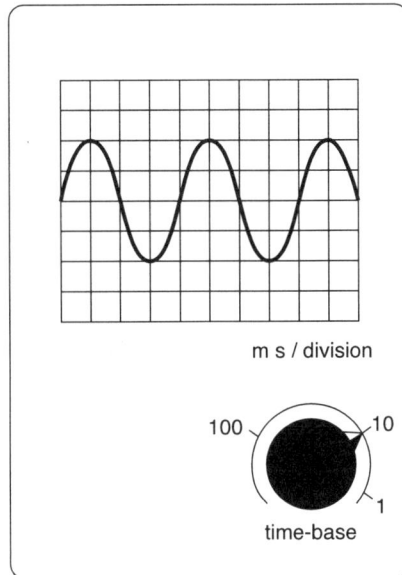

m s / division

time-base

From the oscilloscope picture:

1 wave	$\longrightarrow$	4 divisions
Time for 1 wave	$\longrightarrow$	4 divisions $\times$ 10 m s / division
Time for 1 wave (T)	$\longrightarrow$	40 m s

$$\text{Frequency} \quad = \quad \frac{1}{T} \quad = \quad \frac{1}{40 \times 10^{-3}} \quad = \underline{\underline{25 \text{ Hz}}}$$

Progress to P.P. 'H' P. Page 81, nos 1–5

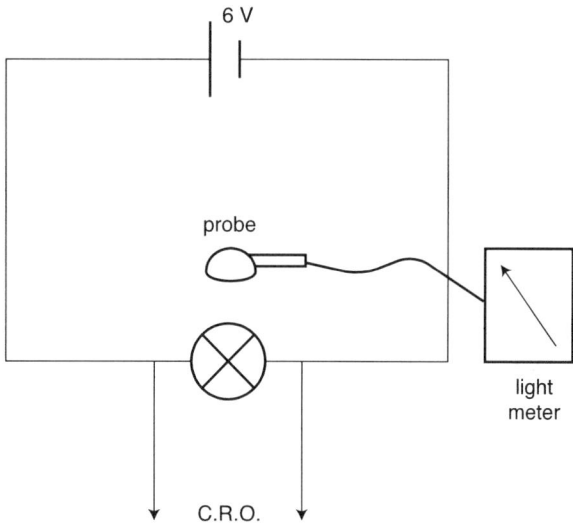

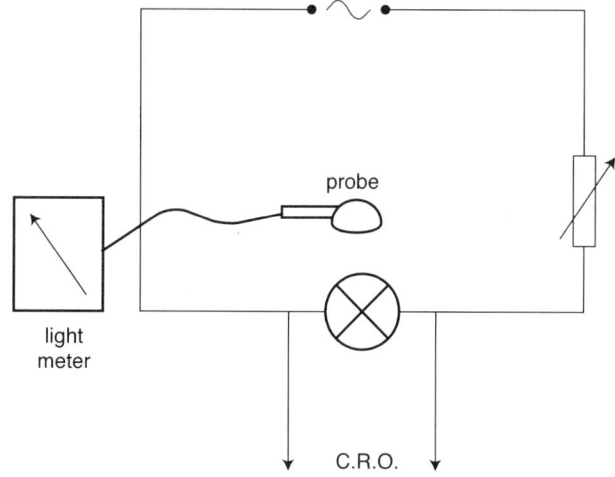

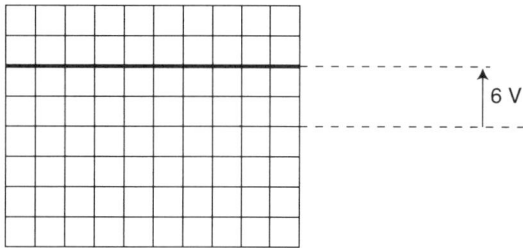

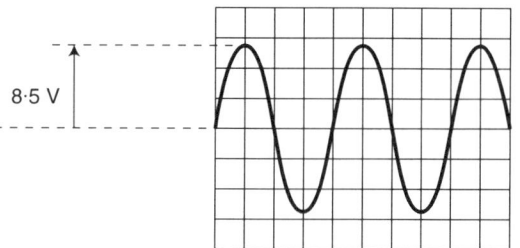

PEAK AND R.M.S. VOLTAGE

A d.c. supply can be used to light a bulb. An a.c. supply can do the same job. Is 6 V d.c. the same as 6 V a.c.?

The experiment opposite gives a comparison of light level from a bulb operated firstly from a d.c. supply and then an a.c. supply.

♦   First, the d.c. supply is connected to the bulb and a reading is taken from the lightmeter.

♦   Second, the a.c. supply is connected to the bulb with a rheostat used to control the brightness of the bulb.

♦   Next the brightness of the bulb is varied to make it the **same** as in the d.c. circuit, i.e., the effect of the d.c. voltage is the same as the effect of the a.c. voltage.

The C.R.O. (cathode ray oscilloscope) is used to measure the voltage in each circuit.

d.c. circuit   voltage measured = <u>6 V</u>
a.c. circuit peak voltage measured = <u>8·5 V</u>

The peak voltage is found to be $\sqrt{2}$ times the d.c. voltage (called r.m.s.).

$$V_{peak} = \sqrt{2}\ V_{r.m.s.}$$

The quoted value of the mains supply in the U.K. is 230 V a.c. This is the r.m.s. value (or the d.c. equivalent value) of the mains supply.

The peak value of the mains supply is

$V_{peak} = \sqrt{2}\ V_{r.m.s.}$
$V_{peak} = \sqrt{2} \times 230$
$V_{peak} = \underline{\underline{325 \cdot 27\ V}}$

Progress to P.P. 'H' P. Page 82, nos 1– 4

LEARN
$V_{peak} = \sqrt{2}\ V_{r.m.s.}$ and $I_{peak} = \sqrt{2}\ I_{r.m.s.}$

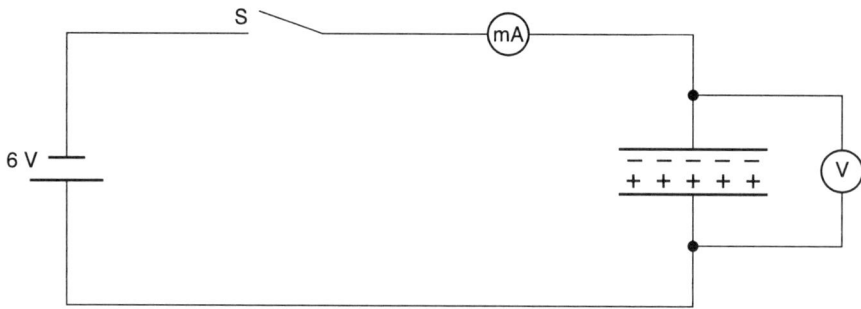

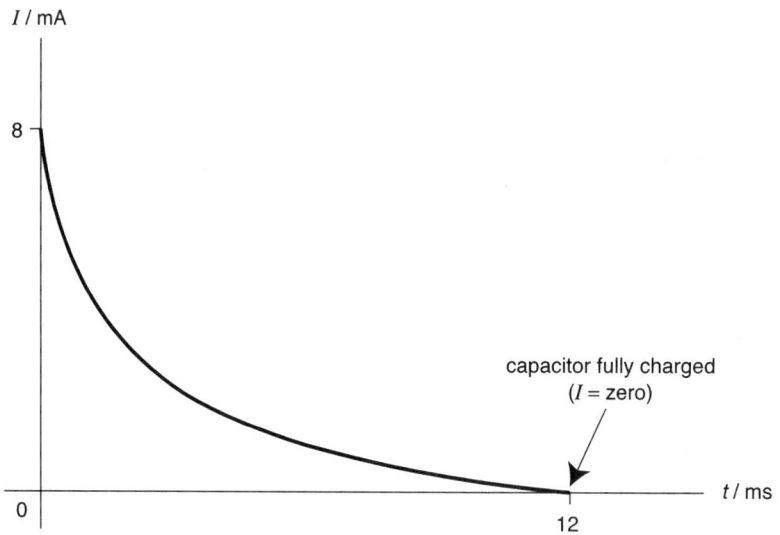

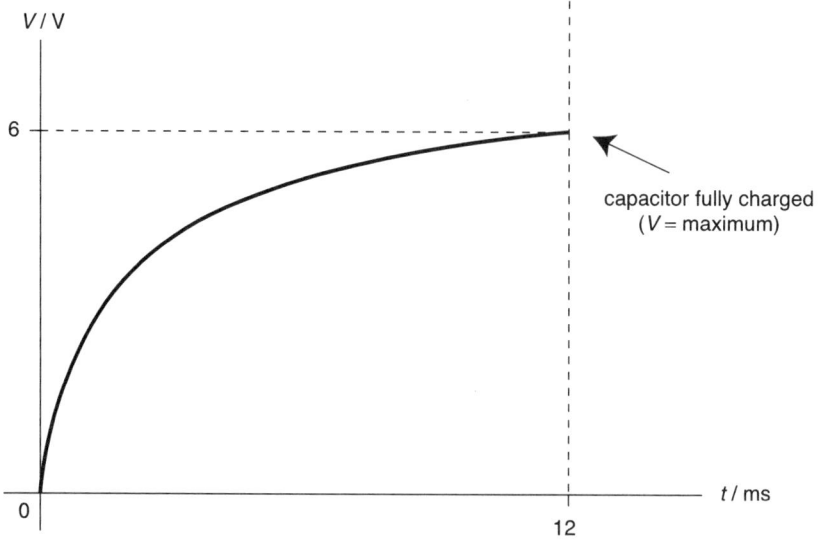

CHAPTER 9

CAPACITANCE

CHARGING A CAPACITOR

When S is closed, electrons flow from the battery to the top plate of the capacitor. This induces an equal and opposite charge on the bottom plate. The first electron on is very easy. The second electron must overcome the repulsion of the first electron. It is even harder for the third electron, etc. The capacitor is full when the force pushing the electrons on is balanced by the force of repulsion.

When the switch is closed, initially the current is high but it quickly decreases to zero as the capacitor charges up.

When fully charged, current is zero

When the switch is closed, initially the voltage is zero but it quickly increases to a maximum (6 V) as the capacitor charges up.

When fully charged, voltage is a maximum (equal to supply voltage)

LEARN
The capacitor is fully charged when • current is zero • voltage is maximum

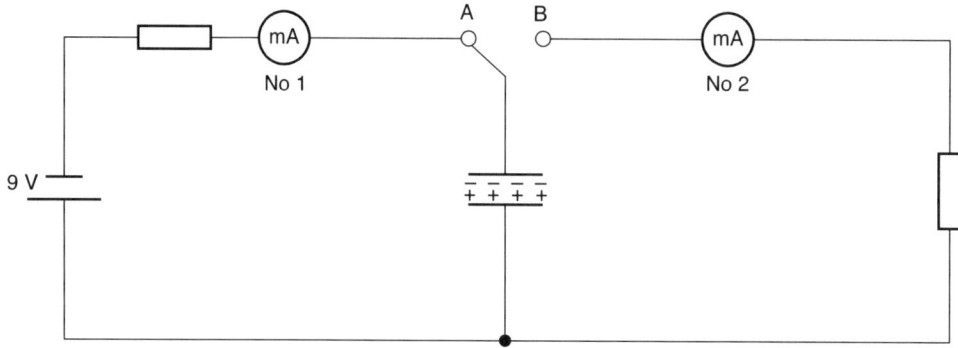

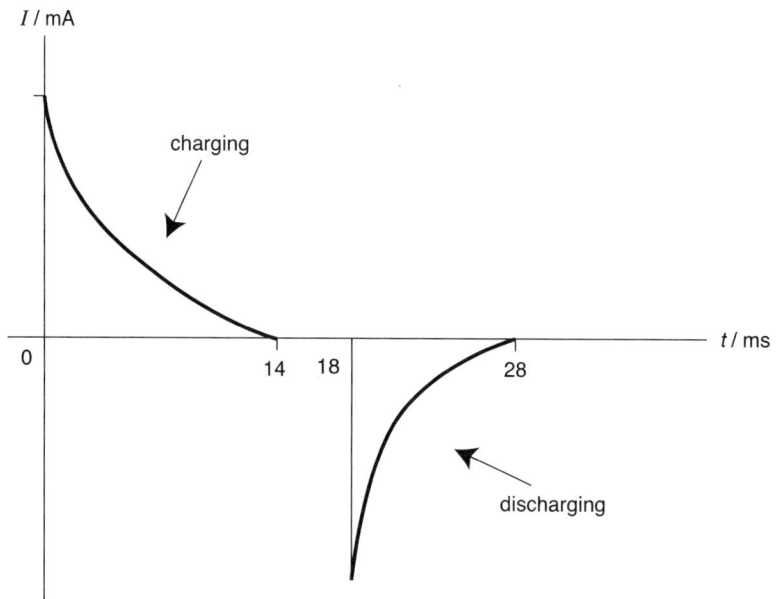

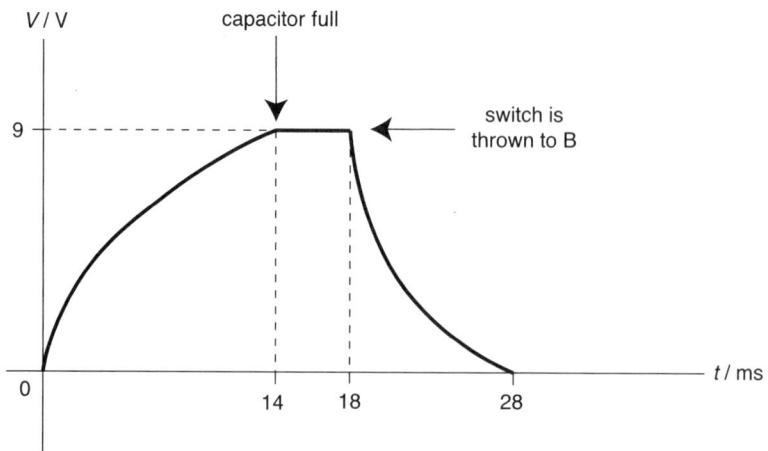

CHARGE AND DISCHARGE OF A CAPACITOR

Circuit

With the switch at A, electrons flow from the battery on to the top plate of the capacitor (the capacitor charges up).

When the switch is thrown to B, electrons on the top plate flow up to B, through milliammeter No 2, through the resistor to the bottom plate of the capacitor (the capacitor discharges).

Current Graph

The capacitor charges up in 14 ms, remains fully charged for 4 ms, at which point the switch is thrown and the capacitor discharges in 10 ms.

Notice the negative current — the discharge current is in the opposite direction to the charging current.

Voltage Graph

The voltage graph is a perfect match for the current graph.

Notice the voltage builds to a maximum (9 V) and during discharge it decreases to zero — but there is no negative voltage.

Progress to P.P. 'H' P. Page 86, nos 1–6

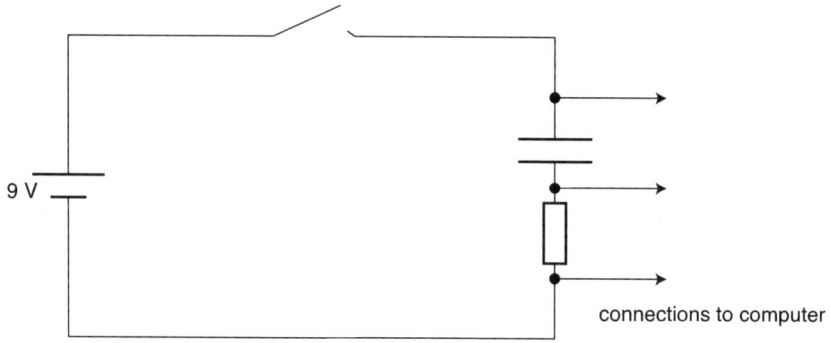

connections to computer

Q / millicoulombs

$$Q \propto V$$

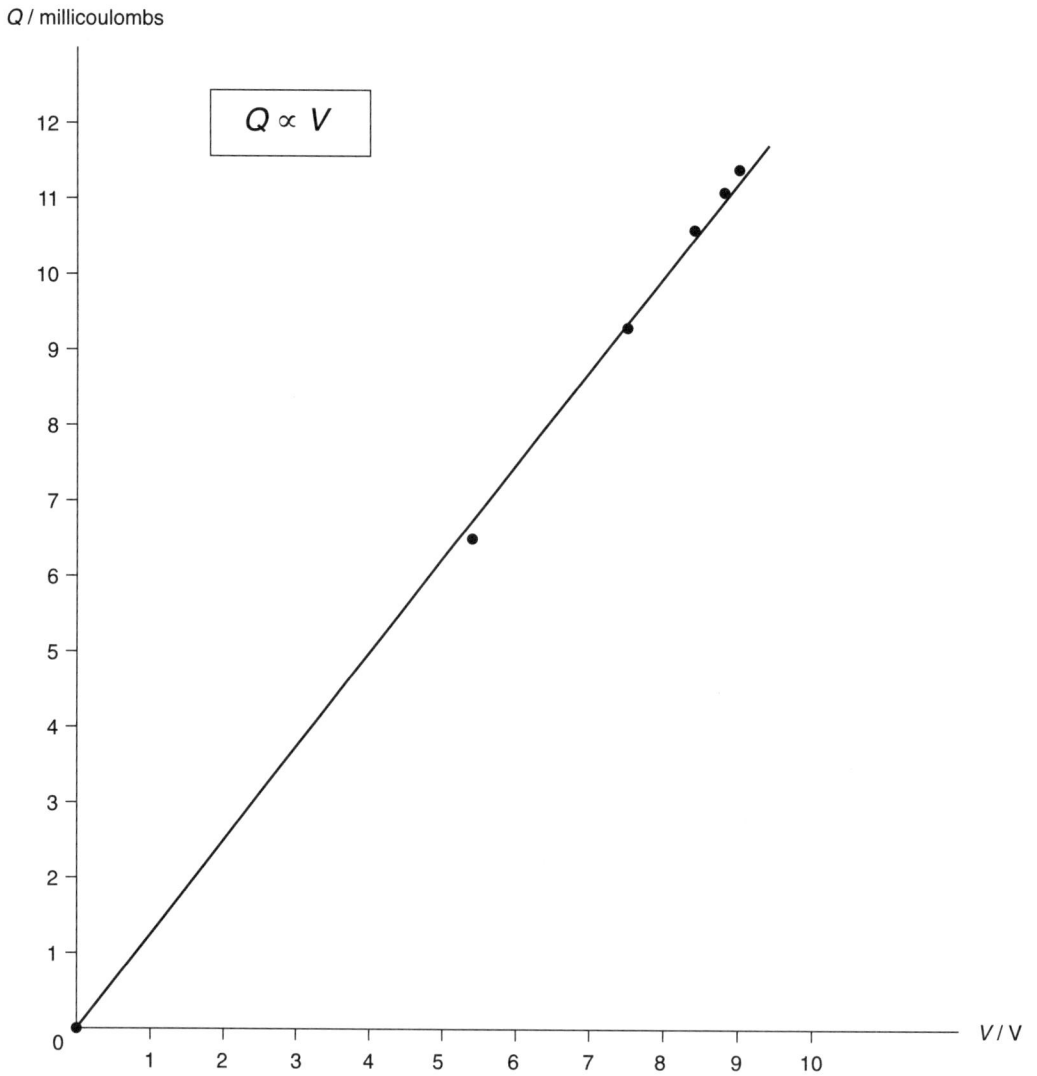

V / V

CHARGE AND VOLTAGE

Investigation (using the computer) to find the relationship between charge and voltage for a capacitor.

When the switch is closed, charge builds up on both plates of the capacitor. The voltage across the capacitor builds up because the charge builds up.

The computer plots the graphs

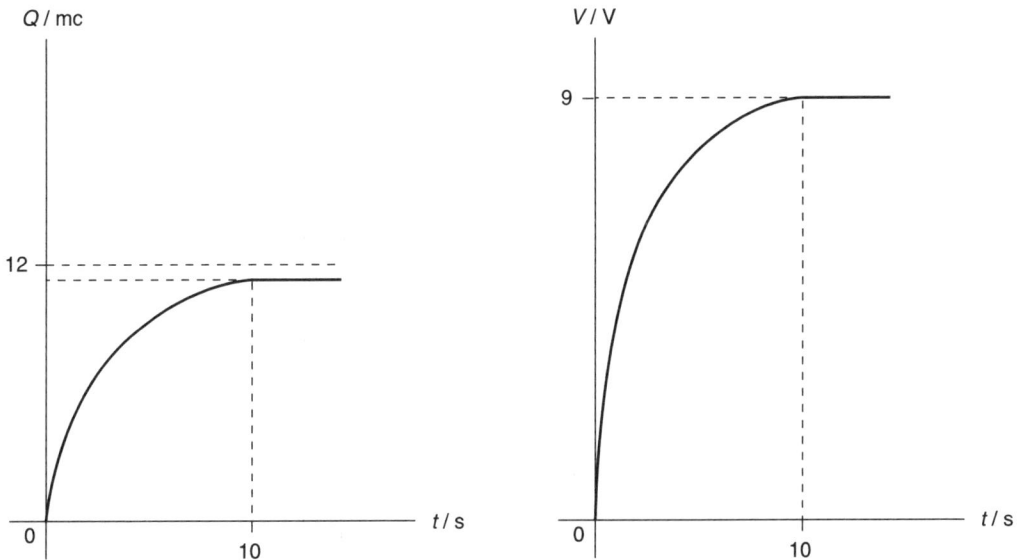

The table of *results* shown is produced by sampling from both graphs above.

This is used to plot a graph of Q against V shown opposite.

V / V	Q / mc
0	0
5·4	6·5
7·5	9·3
8·4	10·6
8·8	11·1
9·0	11·4

Conclusion

$$Q \propto V$$
$$Q = kV \qquad [k = \text{capacitance}]$$
$$\underline{\underline{Q = CV}}$$

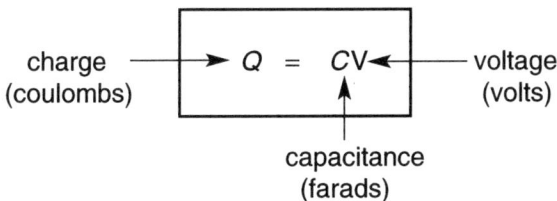

charge (coulombs) ⟶ $Q = CV$ ⟵ voltage (volts)

capacitance (farads)

Progress to P.P. 'H' P. Page 83, nos 1–10

Progress to P.P. 'H' P. Page 88, nos 7–10

LEARN

$$Q = CV$$

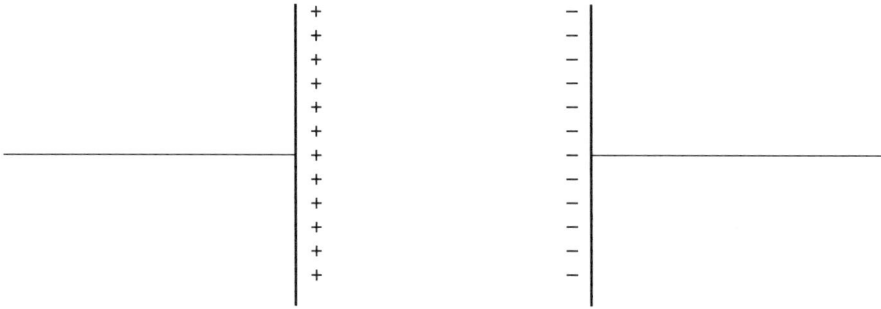

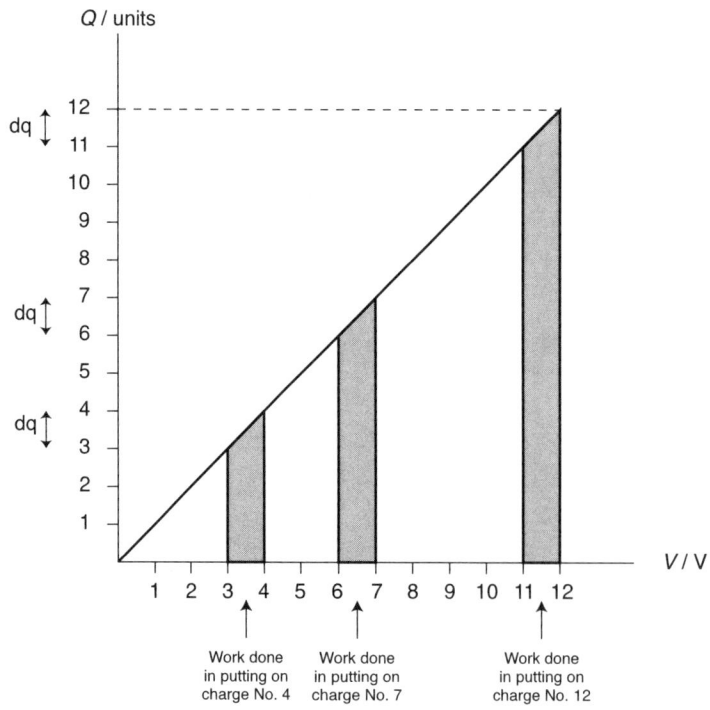

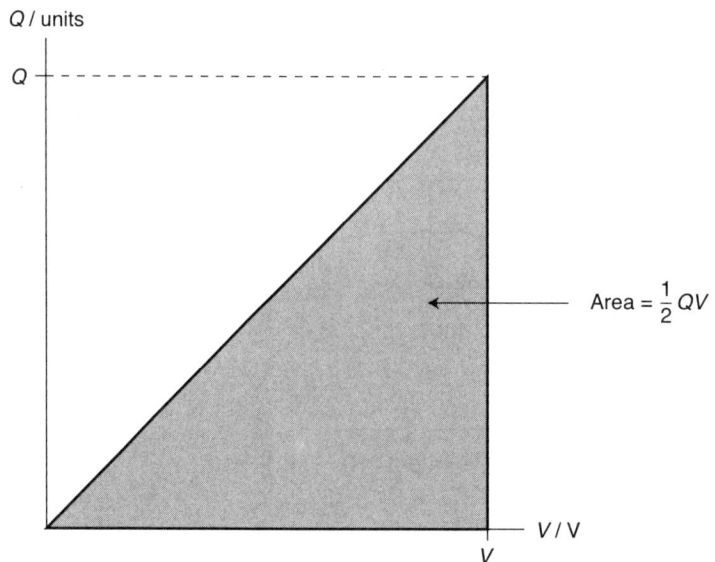

ENERGY STORED IN A CAPACITOR

Imagine the capacitor opposite is charged by 12 equal charges (dq).

Each charge (dq) increases the voltage by 1 V.

Initially the capacitor is uncharged:

> to put on the first charge is very easy;
>
> to put on the second charge is slightly harder;
>
> to put on the third charge is harder still;
>
> to put on the last charge is hardest of all.

It takes energy to put the 4th charge on the capacitor.

It takes more energy to put the 7th charge on the capacitor.

It takes even more energy to put the 12th charge on the capacitor.

The total work done in charging the capacitor is the area under the graph.

$$\text{Energy} = \frac{1}{2}QV$$

Two further equations can be formed by substitution:

$$E = \frac{1}{2}QV \qquad\qquad E = \frac{1}{2}QV$$

$$E = \frac{1}{2}(CV)V \qquad\qquad E = \frac{1}{2}Q\left(\frac{Q}{C}\right)$$

$$E = \frac{1}{2}CV^2 \qquad\qquad E = \frac{Q^2}{2C}$$

$$E = \frac{1}{2}QV = \frac{1}{2}CV^2 = \frac{Q^2}{2C}$$

LEARN
$E = \frac{1}{2}QV$

Progress to P.P. 'H' P. Page 85, nos 1–5

FREQUENCY RESPONSE OF CIRCUIT ELEMENTS

To find out how changing the frequency affects the current ① in a resistor and ② in a capacitor.

①   Resistor

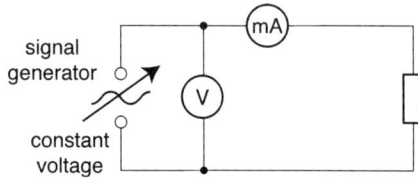

f / Hz	I / mA
0	30
100	30
200	30
300	30
400	30
500	30
600	30
700	30
800	30
900	30
1000	30

In the circuit shown, the frequency is increased and corresponding readings of current are taken (graph opposite).

②   Capacitor

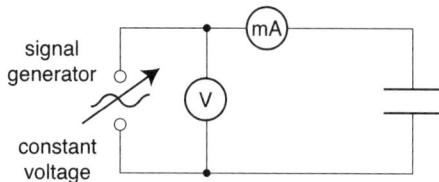

f / Hz	I / mA
0	0
40	10
80	20
120	30
160	40
200	50
240	60
280	70
320	80
360	90

In the circuit shown, the frequency is increased and corresponding readings of current are taken (graph opposite).

I / mA

30

20

10

f / Hz

200   400   600   800   1000

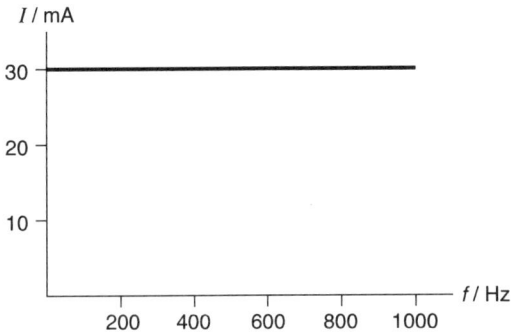

Increasing the frequency has no effect on the current.

Current is independant of frequency.

At 0 Hz (d.c. situation), the current is the same as in the a.c. situation.

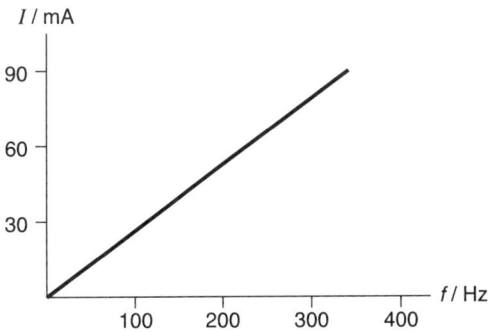

I / mA

90

60

30

f / Hz

100   200   300   400

When the frequency is doubled, the current is doubled.

At 40 Hz, capacitor charges and discharges 40 times in a second (10 mA).

At 80 Hz, capacitor charges and discharges 80 times in a second (20 mA).

current ∝ frequency

At 0 Hz (d.c. situation) the current is zero (capacitor charges up in a fraction of a second and then no more current flows).

The capacitor blocks d.c. but not a.c.

LEARN

Resistor:
    I independant of f
Capacitor:
    $I \propto f$

Progress to P.P. 'H' P. Page 90, nos 1–3

USES OF CAPACITORS

1.   *"Tuning"*

A variable capacitor in a radio can be used to select a single radio station.

2.   *Delay Circuit*

If a delay is required, a capacitor is often used in conjunction with other circuit components, e.g., a transistor. The capacitor takes time to charge up and switch on the transistor which can be linked to a variety of outputs.

3.   *Cross-Over Networks in Loudspeakers*

Capacitors block d.c. but allow a.c. to pass. In fact, high frequency a.c. will pass through a capacitor easier than low frequency a.c.

In the circuit, a mixture of high and low frequency signals are fed to the speakers. The high frequency signals pass easily through the capacitor.

Therefore, high frequencies are directed to loudspeaker 2.

Therefore, low frequencies are directed to loudspeaker 1.

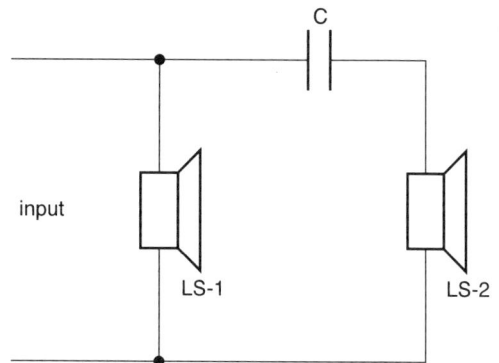

Half-wave rectification using a semiconductor diode

The p–n junction can be used to rectify an a.c. signal.

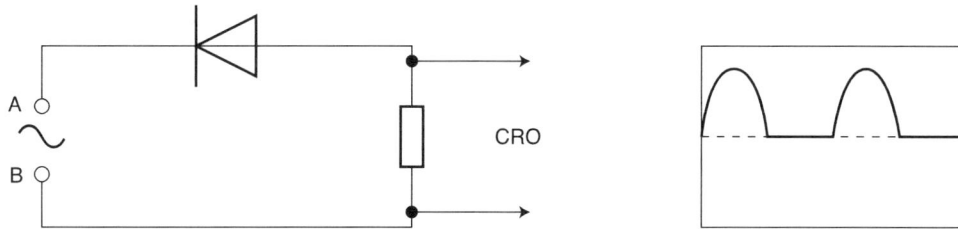

In the circuit, electrons flow from A down through the resistor and back to B in the first half cycle. In the second half cycle there is no current.

4. *Smoothing*

Half wave rectified patterns are "bumpy" and not smooth the way a true d.c. signal is. Rectified signals can be smoothed to resemble d.c. signals more closely by connecting a capacitor in parallel with the resistor across which the rectified signal is taken.

In the circuit below, the capacitor charges up (for polarity see diagram) during the first half cycle when a current flows and then discharges through the resistor during the second half cycle when there is no current from the supply.

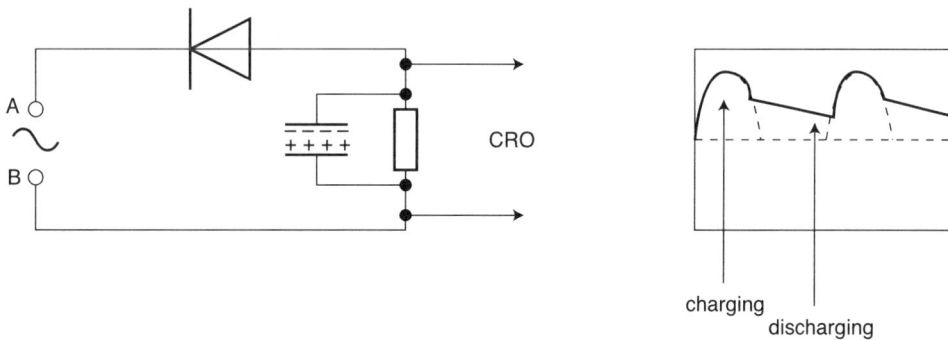

This produces a decreasing discharge current during the second half cycle which covers or "smooths" the gap until the next pulse appears as shown on the oscilloscope wave pattern.

Progress to P.P. 'H' P. Page 91, nos 4–5

CHAPTER 10

ANALOGUE ELECTRONICS

THE OPERATIONAL AMPLIFIER

The operational amplifier is a chip which amplifies but it performs other operations too, like adding or subtracting.

The op-amp has two input terminals and one output terminal.

The inputs are called the inverting input (V_1) and the non-inverting input (V_2).

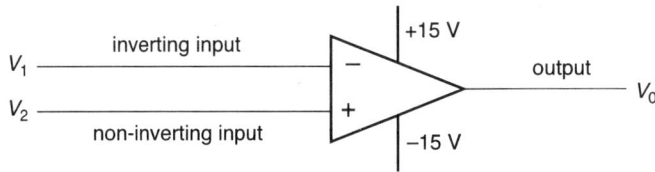

The symbol above includes the positive and negative power supply (usually 15 V) but it is common practice to omit the power supply to achieve a simpler symbol (below).

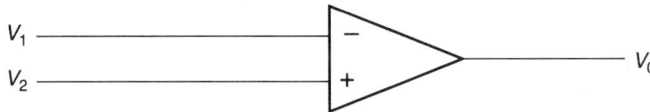

The Ideal Operational Amplifier

An ideal op-amp has:

1.   zero input current, i.e., it has infinite input resistance;

2.   no potential difference between the inverting and non-inverting inputs (V_1 and V_2), i.e., both input pins are at the same potential.

Gain of the Operational Amplifier

The gain of the op-amp is defined as the ratio $\dfrac{V_{out}}{V_{in}}$

$$gain = \frac{V_{out}}{V_{in}}$$

However, this means different things in different circuits.

The op-amp will be used in two modes.

1.   *Inverting Mode*

$$gain = \frac{V_0}{V_1}$$

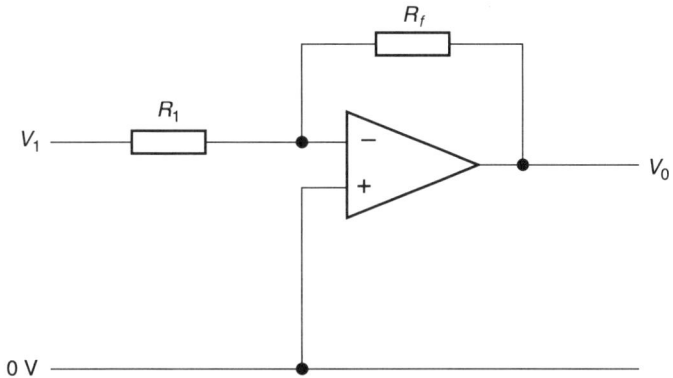

2.   *Differential Mode*

$$gain = \frac{V_0}{(V_2 - V_1)}$$

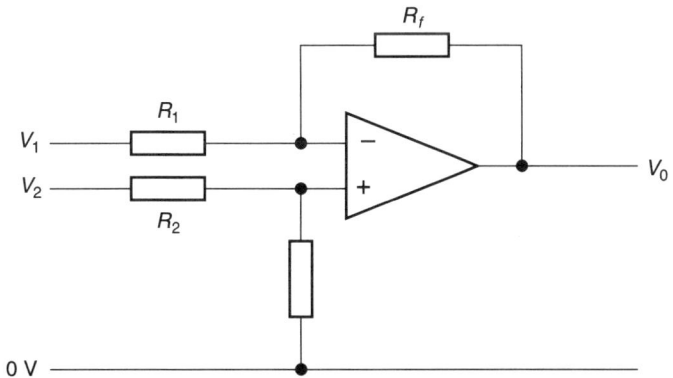

Both circuits use negative feedback, which reduces the gain but makes the gain constant over a large range of frequencies.

N.B. It is useful to revise voltage division before moving on to the inverting mode.

Progress to P.P. 'H' P. Page 92, nos 1–10

LEARN
$gain = \dfrac{V_{out}}{V_{in}}$

INVERTING MODE

The operational amplifier can be used in the inverting mode in the following circuit. Only one input is used (V_1) and the second input (V_2) is connected to earth (0 V).

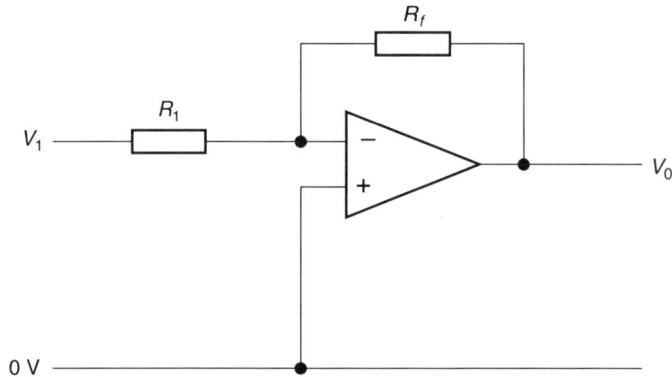

An input signal V_1 will be amplified in the ratio $\dfrac{R_f}{R_1}$ to give V_0 (but the V_0 is the negative of the V_1).

The circuit behaves according to the equation:

$$\frac{V_0}{V_1} = \frac{-R_f}{R_1}$$

EXAMPLE 1

Problem

Calculate V_0 in the circuit shown.

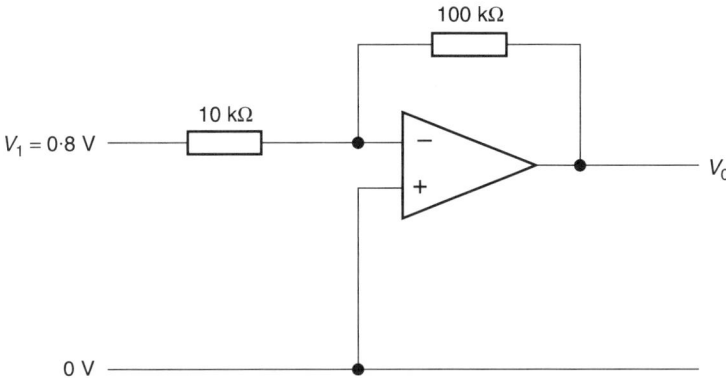

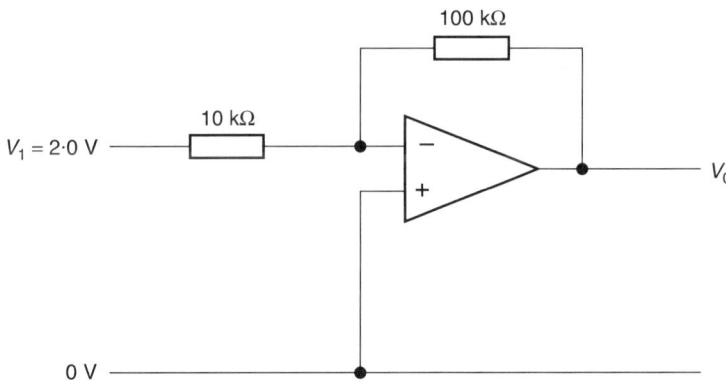

```
                    100 kΩ
            ┌────[        ]────┐
            │                  │
   10 kΩ    │   ╲─            │
V₁ = 0·8 V ─[   ]──●──│ −  ╲   │
                   │  │     ╲──●────── V₀
                   │  │ +   ╱
                   └──│    ╱
                      ╲  ╱

   0 V ──────────────●──────────────────
```

Solution

$$\frac{V_0}{V_1} = \frac{-R_f}{R_1}$$

$$\frac{V_0}{0\cdot8} = -\frac{100\ k\Omega}{10\ k\Omega}$$

$$V_0 = -10 \times 0\cdot8$$

$$\underline{\underline{V_0 = -8\ V}}$$

EXAMPLE 2

Problem

Calculate V_0 in the circuit shown.

```
                    100 kΩ
            ┌────[        ]────┐
            │                  │
   10 kΩ    │   ╲─            │
V₁ = 2·0 V ─[   ]──●──│ −  ╲   │
                   │  │     ╲──●────── V₀
                   │  │ +   ╱
                   └──│    ╱
                      ╲  ╱

   0 V ──────────────●──────────────────
```

Solution

$$\frac{V_0}{V_1} = \frac{-R_f}{R_1}$$

$$\frac{V_0}{2\cdot0} = \frac{-100\ k\Omega}{10\ k\Omega}$$

$$V_0 = -10 \times 2\cdot0$$

$$\underline{\underline{V_0 = -20\ V}}$$

According to the equation, the output voltage should be −20 V. In reality it cannot exceed 85% of its working voltage (15 V), i.e., 12·75 V.

Therefore output here is −12·75 V (saturation).

LEARN
Inverting mode $$\frac{V_0}{V_1} = \frac{-R_f}{R_1}$$ Saturation occurs above 12·75 V

Progress to P.P. 'H' P. Page 96, nos 1–5

EXAMPLE 3

Problem

The circuit below combines a voltage divider with an op-amp in inverting mode. Find V_0.

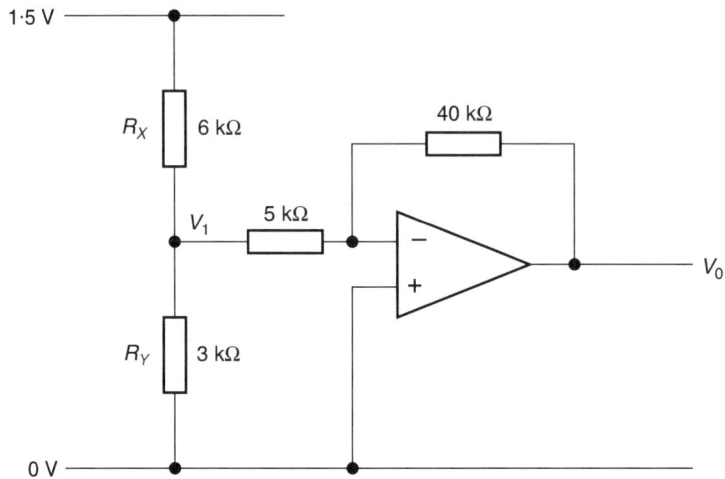

Solution

<table>
<tr><td>First find V_1</td><td>Then find V_0</td></tr>
</table>

First find V_1

$$V_1 = \frac{R_Y}{R_X + R_Y} \times 1.5 \text{ V}$$

$$V_1 = \frac{3 \text{ k}\Omega}{9 \text{ k}\Omega} \times 1.5 \text{ V}$$

$$V_1 = \underline{0.5 \text{ V}}$$

Then find V_0

$$V_0 = \frac{-R_f}{R_1} V_1$$

$$V_0 = \frac{-40 \text{ k}\Omega}{5 \text{ k}\Omega} \times 0.5$$

$$V_0 = \underline{\underline{-4 \text{ V}}}$$

Progress to P.P. 'H' P. Page 97, nos 6–10

Adding Two Voltages

One application of the inverting mode is to add two voltages together. The circuit below adds V_A and V_B.

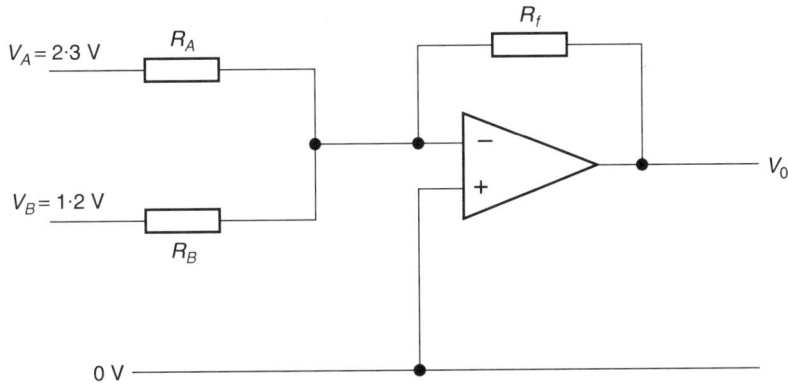

Provided the three resistors, R_A, R_B and R_f, are the same size

$$V_0 = -(V_A + V_B)$$

e.g., $V_0 = -(2 \cdot 3 + 1 \cdot 2) = \underline{\underline{-3 \cdot 5 \text{ V}}}$

i.e., the size of V_0 is the same as the size of $V_A + V_B$.

If V_B is made negative the circuit subtracts.

LEARN
Adding circuit $V_0 = -(V_A + V_B)$

DIFFERENTIAL MODE

The operational amplifier can be used in the differential mode in the following circuit to compare the two input voltages V_1 and V_2.

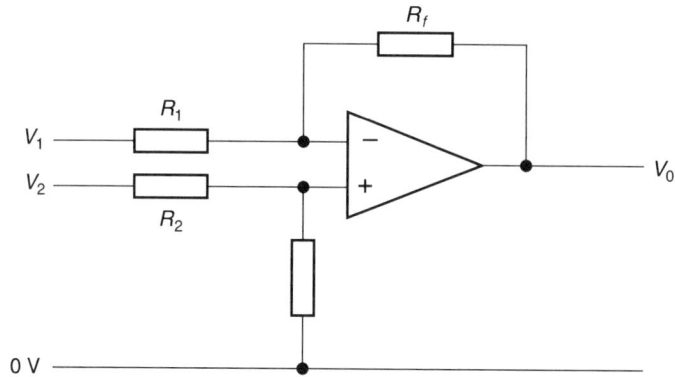

In this mode, the circuit takes the difference between V_2 and V_1, i.e. $(V_2 - V_1)$, and amplifies it in the ratio $\dfrac{R_f}{R_1}$.

The circuit behaves according to the equation

$$V_0 = (V_2 - V_1)\ \frac{R_f}{R_1}$$

When V_2 and V_1 are the same, there is no difference and so nothing to amplify. Hence the output voltage V_0 is 0 V.

EXAMPLE 1

Problem

Calculate V_0 in the circuit shown.

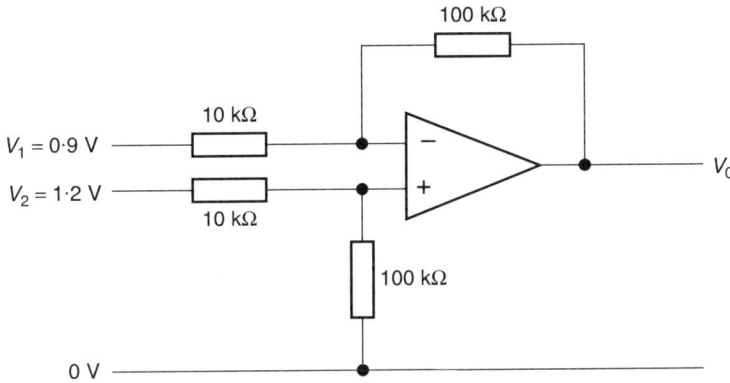

Solution

$$V_0 = (V_2 - V_1)\frac{R_f}{R_1}$$

$$V_0 = (1\cdot2 - 0\cdot9)\frac{100\ k\Omega}{10\ k\Omega}$$

$$V_0 = \underline{\underline{3\ V}}$$

EXAMPLE 2

Problem

The circuit below combines a Wheatstone Bridge circuit with an op-amp in differential mode. Find V_0.

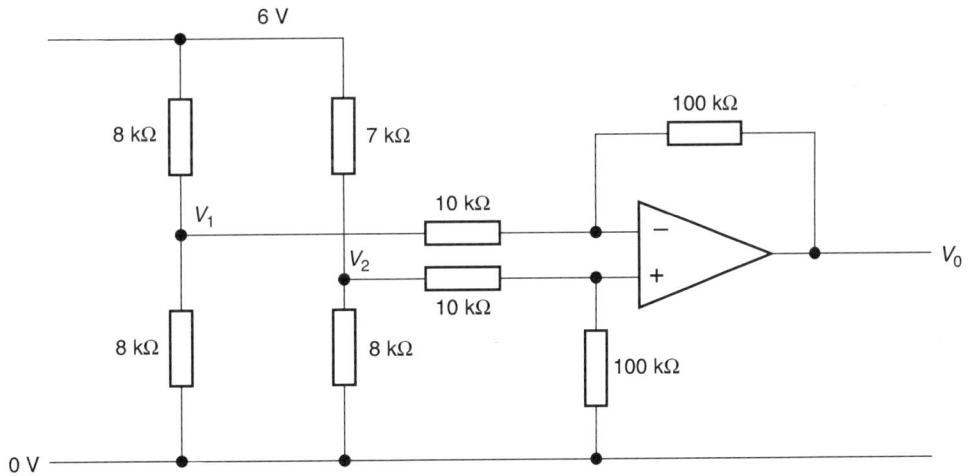

Solution

First find V_1

$$V_1 = \frac{8\ k\Omega}{16\ k\Omega} \times 6\ V$$
$$V_1 = \underline{3\ V}$$

Next find V_2

$$V_2 = \frac{8\ k\Omega}{15\ k\Omega} \times 6\ V$$
$$V_2 = \underline{3\cdot2\ V}$$

Finally find V_0

$$V_0 = (V_2 - V_1)\frac{R_f}{R_1}$$
$$V_0 = (3\cdot2 - 3)\frac{100\ k\Omega}{10\ k\Omega}$$
$$V_0 = \underline{\underline{2\ V}}$$

LEARN

Differential mode

$$V_0 = (V_2 - V_1)\frac{R_f}{R_1}$$

Progress to P.P. 'H' P. Page 99, nos 1–10

CHAPTER 11

WAVES

PARTS OF A WAVE

The **wavelength** (λ) of a wave is the length of one wave. It is also the distance between successive crests (or troughs).

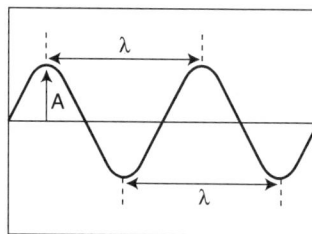

The **amplitude (A)** of a wave is the height of the crest above the centre. The greater the amplitude of a wave, the greater the energy of the wave.

The **frequency (f)** is the number of waves per second. The frequency of the wave motion is the same as the frequency of the source producing it, e.g., tuning fork.

The **period (T)** is the time taken for one cycle. If the frequency of a source is 50 Hz, the period is $\frac{1}{50}$ second.

Hz

$$\therefore f = \frac{1}{T}$$

s

The **wave equation**

Wave speed can be calculated from:

m s^{-1}

$$v = f\lambda$$

m

Hz

The Four Phenomena of Waves

All types of waves exhibit the characteristic behaviour of reflection, refraction, diffraction and interference.

Reflection

barrier

water waves

$i \quad r$

N

$\angle i = \angle r$

Refraction

Deep (fast)   Shallow (slow)

N

Water waves slow down when they enter shallow region and bend towards normal. Wavelength gets shorter.

Diffraction

Gap similar to λ.

Interference

Interference by division of wavefront. Both sources are coherent.

INTERFERENCE OF WATER WAVES

When one water wave meets another identical water wave, the two waves interfere.

When the crest of one wave meets the crest of another, the two waves interfere constructively to produce a crest double the height.
This is called constructive interference.

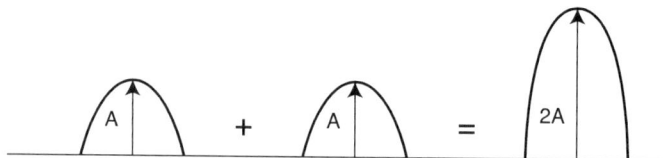

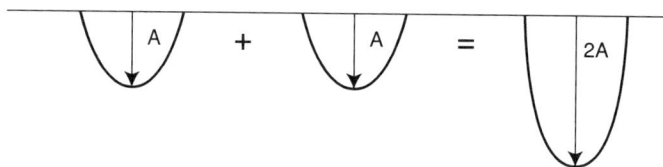

When the trough of one wave meets the trough of another, the two waves interfere constructively to produce a trough double the depth.
This is called constructive interference.

When the crest of one wave meets the trough of another, the two waves interfere destructively to produce flat or calm water.
This is called destructive interference.

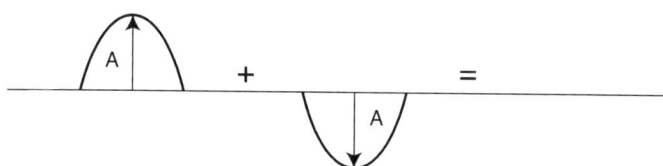

In the ripple tank, we can produce two identical waves together by:

(a) using two dippers vibrating up and down perfectly in time or *in phase* with each other,

OR

(b) using a barrier with two gaps so that plane waves incident on the gaps produce a diffraction pattern at each gap, i.e., a double diffraction pattern. The two diffraction patterns overlap to produce an interference pattern.

Choosing the second method, the interference pattern shown opposite is formed.

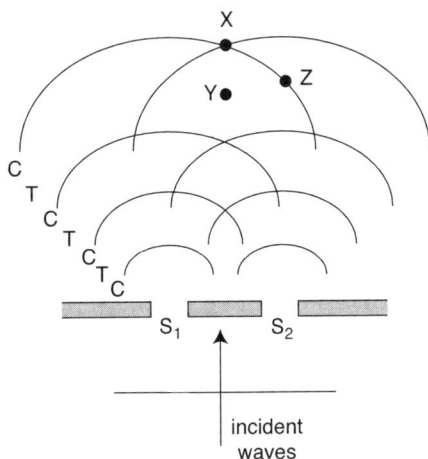

C = crest
T = trough

At point X, a crest from source S_1 meets a crest from source S_2 and they add together to produce a crest double the normal height — constructive interference.

At point Y, a trough from source S_1 meets a trough from source S_2 and they add together to produce a trough double the depth — constructive interference.

At point Z, a crest from source S_1 meets a trough from source S_2 and they cancel each other out to produce calm water — destructive interference.

These constructive and destructive interference points form alternate lines spreading out from the two sources.

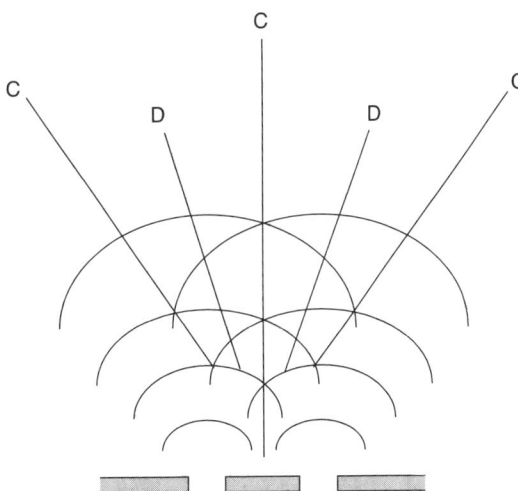

C = constructive interference line
D = destructive interference line

It is like stretched fingers spreading out from a hand where the fingers represent constructive bands and the spaces between them represent destructive bands.

Progress to P.P. 'H' P. Page 103, nos 1–4

INTERFERENCE OF LIGHT

In 1801, Young performed an experiment to prove that light was a wave motion. Previously it had been thought that light was made of tiny particles.

Young argued that if light is a wave motion it will exhibit the **phenomenon** of interference. He knew water waves would diffract through a gap and if he used two gaps he could achieve double diffraction and interference where the waves overlap.

Young adapted this idea for light waves. He thought that if light was a wave motion, it must have a very small wavelength. With two tiny gaps he could produce diffraction of the light at each gap. If he was successful, he would see interference between the two gaps.

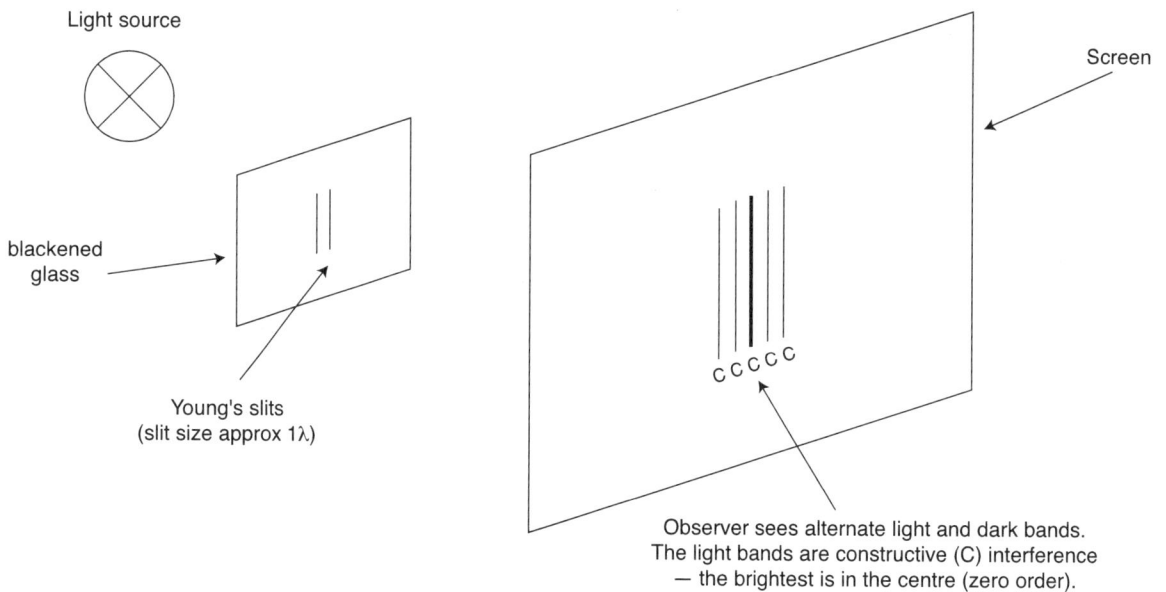

Light source

Screen

blackened
glass

Young's slits
(slit size approx 1λ)

C C C C C

Observer sees alternate light and dark bands.
The light bands are constructive (C) interference
— the brightest is in the centre (zero order).

An interference pattern would show alternate bright and dark bands. There will be an odd number of the bright bands (constructive interference). The dark bands are areas of destructive interference.

Young was successful in proving light was a wave because he was able to show interference (the test for a wave).

EQUATIONS

The diagram below shows how the first order maximum is produced in an interference pattern using monochromatic (single λ) light.

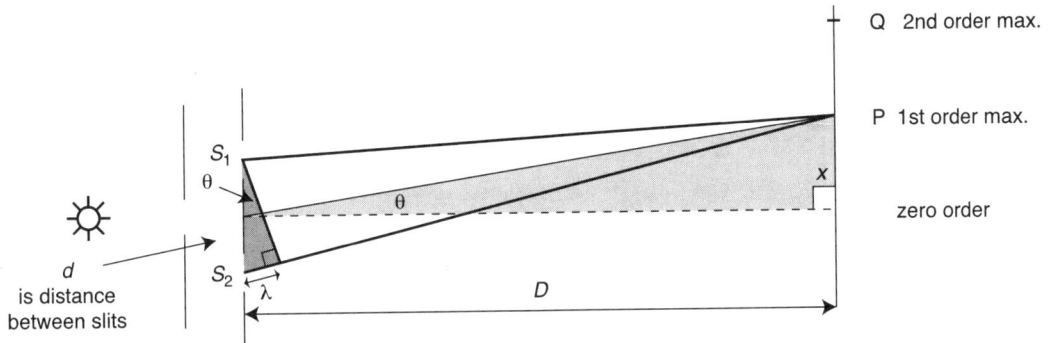

The first order maximum occurs the first time (from zero order) light from S_1 and S_2 are in phase, i.e., $\boxed{S_2P - S_1P = \lambda}$.

The second order maximum occurs the second time (from zero order) light from S_1 and S_2 are in phase, i.e., $\boxed{S_2Q - S_1Q = 2\lambda}$.

Both shaded triangles above are similar.

From small triangle

$$\sin \theta = \frac{\lambda}{d}$$

> note for 2nd order, $\sin \theta = \dfrac{2\lambda}{d}$
>
> generally $n\lambda = d \sin \theta$

From large triangle

$$\tan \theta = \frac{x}{D}$$

Combining:

for small θ: $\sin \theta = \tan \theta$

$$\therefore \frac{\lambda}{d} = \frac{x}{D}$$

$$\boxed{\therefore \lambda = \frac{xd}{D}}$$

d = slit separation,
x = fringe separation,
D = distance from the slits to the screen.

Factors affecting fringe separation x

$$x = \frac{\lambda D}{d}$$

$\therefore x \propto \lambda$

$\therefore x \propto D$

$\therefore x \propto \dfrac{1}{d}$

LEARN
Path difference = $n\lambda$ (max)
Path difference = $\left(n + \dfrac{1}{2}\right)\lambda$ (min)
$n\lambda = d \sin \theta$

Progress to P.P. 'H' P. Page 105, nos 5–9

EXPERIMENT TO FIND THE WAVELENGTH OF THE LASER LIGHT

The diagram shows light from a laser striking a diffraction grating.

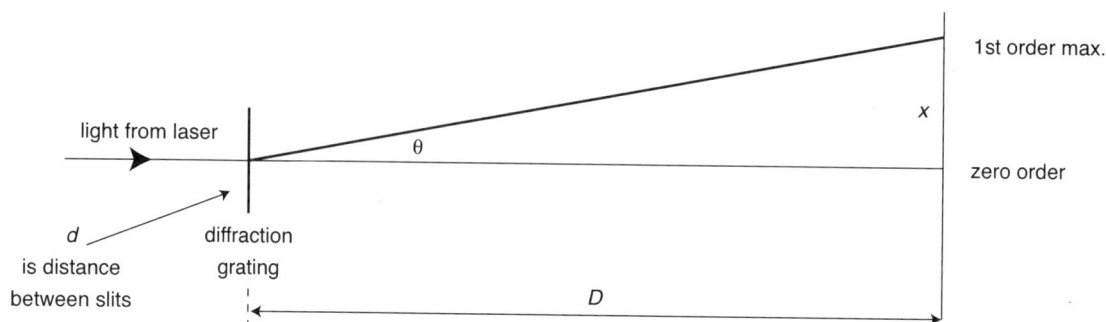

Results

Three measurements are taken from the apparatus by a student:

d = slit separation, (600 lines per mm) $= 1 \times 10^{-3}$ m ÷ 600
x = fringe separation, $= 72 \cdot 0$ cm
D = distance from the slits to the screen $=$ $2 \cdot 00$ m

Calculation

$$n \lambda = d \sin \theta$$

$$1 \lambda = \frac{1 \times 10^{-3} \text{ m}}{600} \times \frac{(72 \times 10^{-2} \text{ m})}{2 \cdot 00 \text{ m}}$$ since $\sin \theta = \dfrac{x}{D}$

$$\lambda = 6 \times 10^{-7} \text{ m}$$

$$\lambda = 600 \times 10^{-9} \text{ m}$$

$$\lambda = \underline{\underline{600 \text{ nm}}}$$

Not quite the right answer but within 5% of the actual value.

WAVELENGTHS OF LIGHT

The approximate values for the wavelength of red, green and blue light are:

red   700 nm
green   540 nm
blue   490 nm.

WORKED EXAMPLE ON INTERFERENCE USING MICROWAVES

Problem

The sketch shows the experimental arrangement used to investigate microwaves passing through a double-slit system.

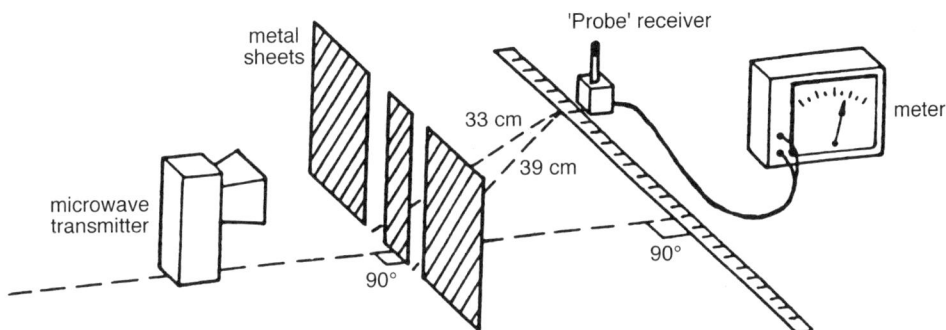

As the probe is moved along the scale, the meter reading increases and decreases repeatedly.

(a) The probe gives one of its maximum readings in the position shown, 33 cm from one slit and 39 cm from the other.

Could the wavelength of the microwaves be
(i)   3 cm,
(ii) 12 cm?

Give a reason for your answer in each case.

(b) Explain how a minimum reading on the meter occurs.

Solution

(a) (i) YES, because the 6 cm difference could be two wavelengths, $\therefore$ path length difference = 6 cm = 2λ, i.e., the probe is at the 2nd order maximum.

(ii) NO, because the 6 cm difference could only be half a wavelength, $\therefore$ path length difference = 6 cm = $\frac{1}{2}\lambda$, i.e., the probe would be at the first minimum from the zero order.

(b) Minimum is caused by destructive interference, i.e., a crest from one slit meets a trough from the other slit and cancel each other out.

Progress to P.P. 'H' P. Page 108, nos 10–12

PRISMS AND GRATINGS

Refraction of Monochromatic Light

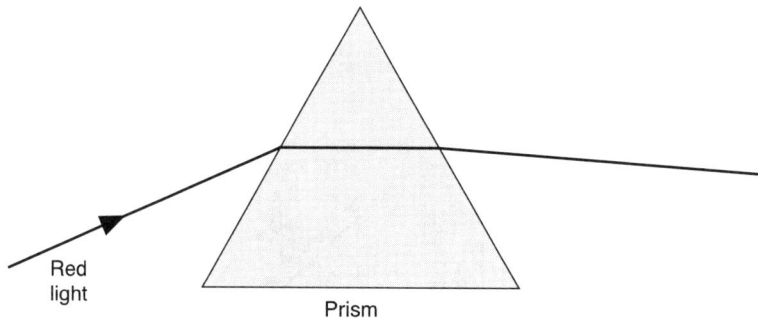

Red
light

Prism

Monochromatic light slows down inside the glass but will not split into different colours because there is only one wavelength.

Refraction of White Light

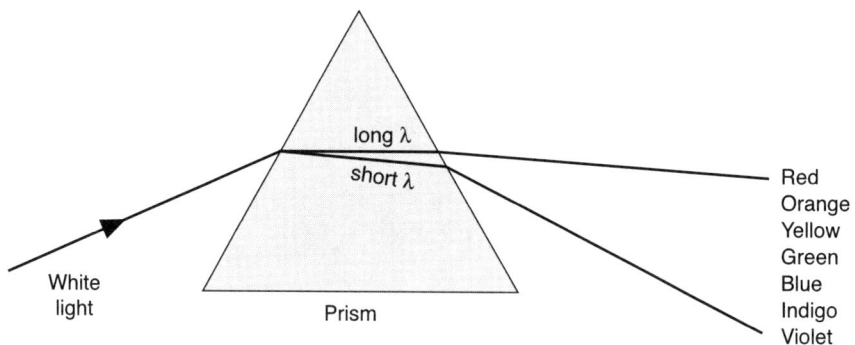

long λ

short λ

White
light

Prism

Red
Orange
Yellow
Green
Blue
Indigo
Violet

White light slows down and splits into the colours of the spectrum because longer wavelengths (red) are slowed down less than shorter wavelengths (violet).

RED IS REFRACTED LESS THAN VIOLET

Diffraction of Monochromatic Light

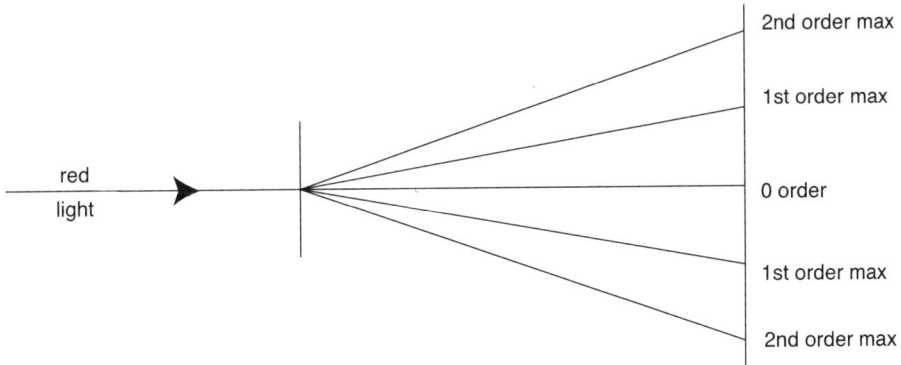

red
light

2nd order max

1st order max

0 order

1st order max

2nd order max

Monochromatic light gives an interference pattern which is also monochromatic.

Diffraction of White Light

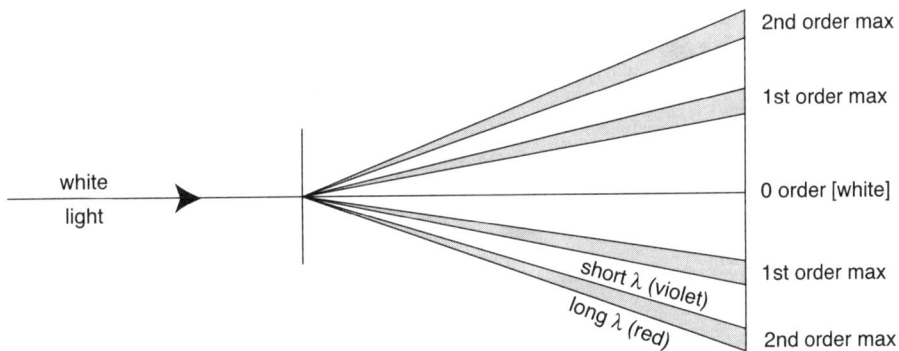

white
light

2nd order max

1st order max

0 order [white]

short λ (violet)
long λ (red)

1st order max

2nd order max

White light gives an interference pattern which shows a spectrum at each (1st, 2nd, etc.) order maximum and white at the zero order because longer wavelengths (red) are diffracted more than shorter wavelengths (violet).

RED IS DIFFRACTED MORE THAN VIOLET

LEARN

LONG λ is diffracted more than SHORT λ

LONG λ is refracted less than SHORT λ

Progress to P.P. 'H' P. Page 109, nos 1–5

Apparatus

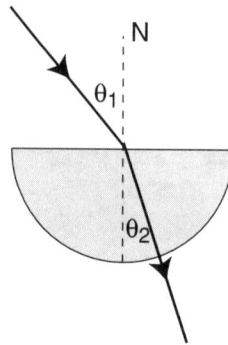

Graph

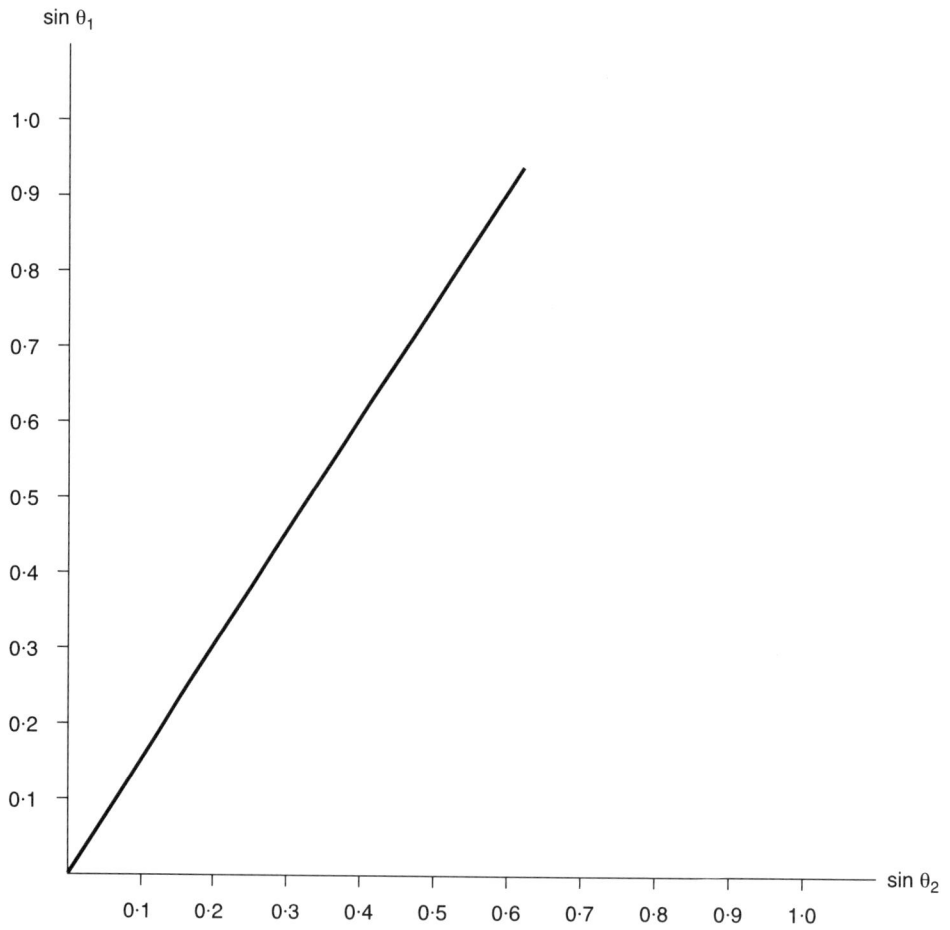

CHAPTER 12

REFRACTION OF LIGHT

REFRACTIVE INDEX

In the diagram opposite, light strikes the glass at an angle of incidence θ_1, slows down and bends towards the normal. The angle of refraction, θ_2, is smaller than the angle of incidence.

Object

To find a relationship between angles θ_1 and θ_2.

Method

The angle of incidence θ_1 is changed and the corresponding angle of refraction θ_2 is measured.

Results

θ_1 / 0	θ_2 / 0	$\sin \theta_1$	$\sin \theta_2$	$\dfrac{\sin \theta_1}{\sin \theta_2}$
10·0	6·5	0·174	0·113	1·54
20·0	13·0	0·342	0·225	1·52
30·0	19·5	0·500	0·334	1·50
40·0	25·5	0·643	0·431	1·49
50·0	31·5	0·766	0·522	1·47
60·0	36·0	0·866	0·588	1·47
70·0	39·0	0·940	0·629	1·49

Conclusion

From the graph
$$\sin \theta_1 \propto \sin \theta_2$$
$$\therefore \quad \sin \theta_1 = k \sin \theta_2 \quad (k = \text{a constant})$$
$$\therefore \quad k = \frac{\sin \theta_1}{\sin \theta_2}$$

The constant is called the refractive index n

$$\therefore \quad n = \frac{\sin \theta_1}{\sin \theta_2} = \frac{\sin \theta_{air}}{\sin \theta_{glass}}$$

LEARN

$$n = \frac{\sin \theta_1}{\sin \theta_2}$$

Progress to P.P. 'H' P. Page 111, nos 1–10

REFRACTIVE INDEX AND VELOCITY

A narrow beam of light passes through a glass block.

When light **enters the glass**, three things happen:

1.   the light slows down;

2.   the light bends towards the normal;

3.   the wavelength (λ) decreases.

When light **leaves the glass**, three things happen:

1.   the light speeds up;

2.   the light bends away from the normal;

3.   the wavelength (λ) increases.

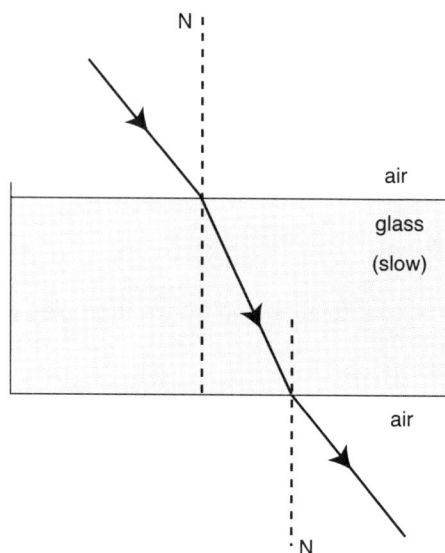

Refractive index can also be calculated from the ratio of velocities, i.e.,

$$n = \frac{\text{velocity of light in air}}{\text{velocity of light in glass}}$$

$$\boxed{n = \frac{v_{air}}{v_{glass}} = \frac{\cancel{f}\,\lambda_{air}}{\cancel{f}\,\lambda_{glass}} = \frac{\lambda_{air}}{\lambda_{glass}}}$$

Notice that the frequency of light does not change.

Example 1

The refractive index of glass is 1·50. How fast does light travel in the glass?

$$n = \frac{v_{air}}{v_{glass}}$$

$$\therefore \quad v_{glass} = \frac{v_{air}}{n}$$

$$v_{glass} = \frac{3 \times 10^8}{1 \cdot 50}$$

$$= \underline{\underline{2 \times 10^8 \text{ m s}^{-1}}}$$

Example 2

White light is made up of many different wavelengths.
One wavelength is 600 nm (in air).
What happens to this wavelength when it enters the glass ($n = 1 \cdot 50$)?

$$n = \frac{\lambda_{air}}{\lambda_{glass}}$$

$$\lambda_{glass} = \frac{\lambda_{air}}{n}$$

$$\lambda_{glass} = \frac{600 \text{ nm}}{1 \cdot 50}$$

$$= \underline{\underline{400 \text{ nm}}}$$

I.e., the wavelength drops from 600 nm to 400 nm.

LEARN

$$n = \frac{v_{air}}{v_{glass}} = \frac{\lambda_{air}}{\lambda_{glass}}$$

Progress to P.P. 'H' P. Page 115, nos 1–10

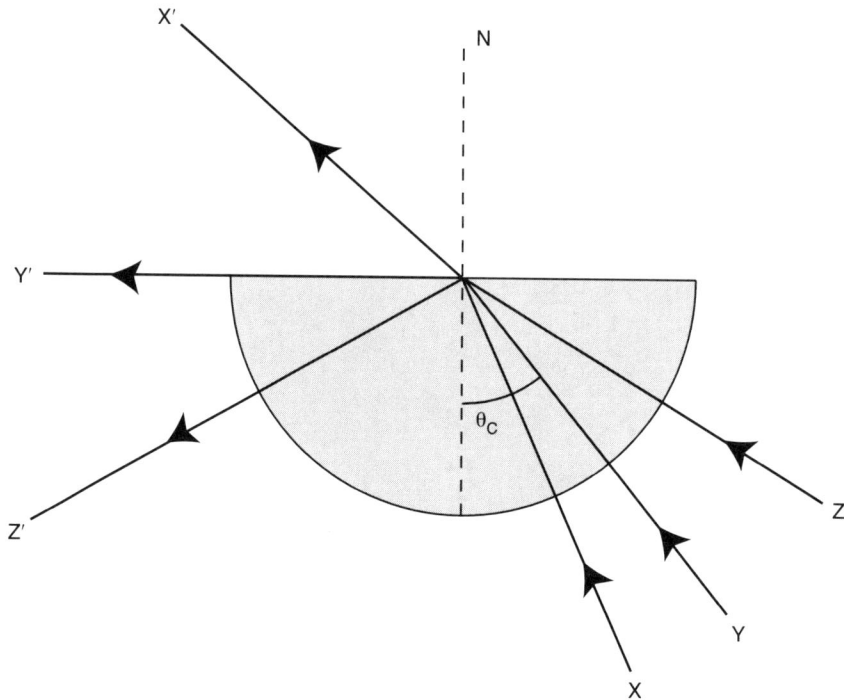

Light only penetrates the boundary if the angle is less than the critical angle (θ_C).

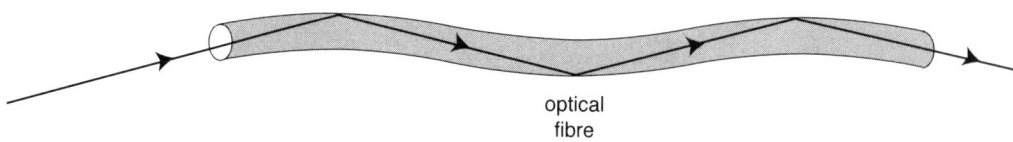

optical
fibre

CRITICAL ANGLE AND TOTAL INTERNAL REFLECTION

Light strikes the glass-air boundary (opposite). Initially the angle between the ray of light and the normal is small, but it is gradually increased from X to Y to Z.

Ray X strikes the boundary at a sharp angle, penetrates the boundary and escapes from the glass (X'). A little light is reflected (not shown).

Ray Y strikes the boundary at the critical angle. The light runs along the surface of the glass at 90° to the normal (Y'). Again some light is reflected (not shown).

Ray Z strikes the boundary at a large angle (greater than the critical angle). Consequently it does not pierce the boundary and all the light is reflected internally (Z'). This is called total internal reflection.

The Critical Angle is the angle of incidence in the glass which produces an angle of refraction of 90°.

Pretend light is entering the glass from Y' to Y (principle of reversibility of light), then $\theta_1 = 90°$ and $\theta_2 = \theta_C$.

$$\therefore \quad n \quad = \quad \frac{\sin \theta_1}{\sin \theta_2}$$

$$\therefore \quad n \quad = \quad \frac{\sin 90}{\sin \theta_C}$$

$$\boxed{\therefore \quad n \quad = \quad \frac{1}{\sin \theta_C}}$$

LEARN
$n = \dfrac{1}{\sin \theta_C}$

Progress to P.P. 'H' P. Page 117, nos 1−6

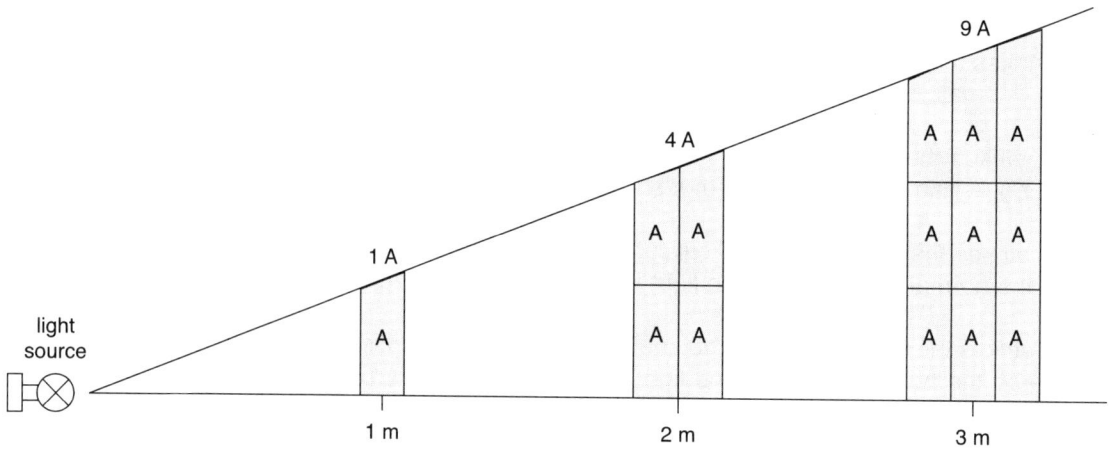

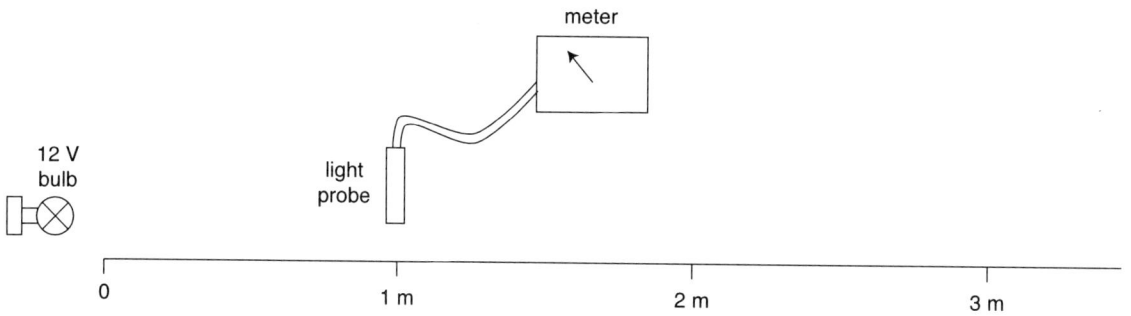

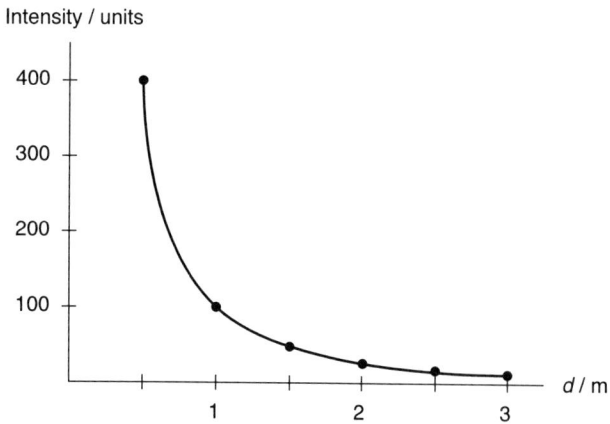

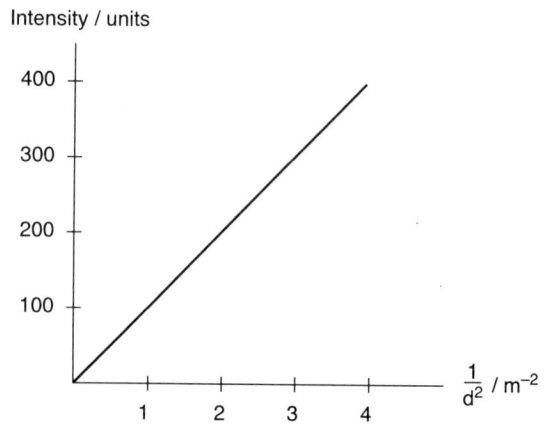

CHAPTER 13

OPTOELECTRONICS AND SEMICONDUCTORS

INTENSITY OF LIGHT

The diagram opposite shows how light from a point source spreads out. As the distance from the source increases, the light covers a larger area.

At 1 m from the source the light covers an area A.
At 2 m from the source the light covers an area 4A.
At 3 m from the source the light covers an area 9A.

Definition

The intensity (I) at the surface on which radiation is incident is the power (P) per unit area (A).

$$I = \frac{P}{A}$$

In the *experiment* shown opposite, a light probe is used to measure the intensity of light at different distances from the light source (in a darkened room).

Results

d / m	$I /$ units	$\frac{1}{d^2} / m^{-2}$
0·5	400	4·00
1·0	100	1·00
1·5	44	0·44
2·0	26	0·25
2·5	17	0·16
3·0	12	0·11

A graph of intensity against $\frac{1}{d^2}$ is a straight line through the origin.

$$\therefore \quad I \propto \frac{1}{d^2}$$
$$\therefore \quad Id^2 = k \; (k = \text{a constant})$$

LEARN
$I = \frac{P}{A}$
$I \propto \frac{1}{d^2}$
$I_1 d^2_1 = I_2 d^2_2$

Progress to P.P. 'H' P. Page 120, nos 1–6

Situation A

tungsten
filament
lamp

white
light

+ + + + zinc

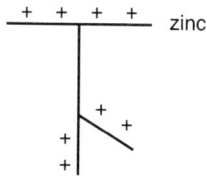

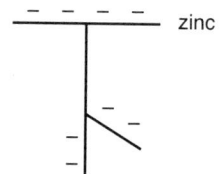

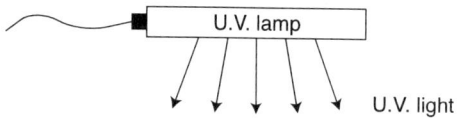

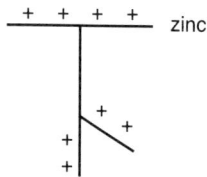

nothing happens

Situation B

tungsten
filament
lamp

white
light

– – – – zinc

nothing happens

Situation C

U.V. lamp

U.V. light

+ + + + zinc

nothing happens

Situation D

U.V. lamp

U.V. light

– – – – zinc

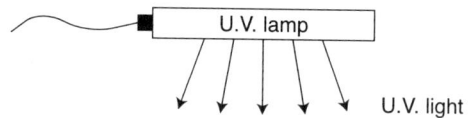

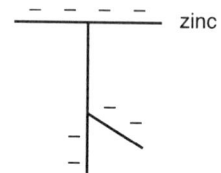

leaf collapses

THE PHOTOELECTRIC EFFECT

Four experimental observations are shown opposite.

A Shining white light on to a positively charged gold leaf electroscope — nothing happens.

B Shining white light on to a negatively charged gold leaf electroscope — nothing happens.

C Shining U.V. light on to a positively charged gold leaf electroscope — nothing happens.

D Shining U.V. light on to a negatively charged gold leaf electroscope — the leaf collapses.

Conclusions

1. Comparing the two bottom boxes, the U.V. light has no effect on positive charges, suggesting that positive charges are heavier than electrons (in agreement with Thomson's model of the atom).

2. Comparing the last box (D) with the box above it (B), the U.V. light has more energy than the white light, causing electrons to escape from the zinc surface.

The electrons require a certain minimum energy to escape. Energy is proportional to the frequency of light according to:

Energy (J) ⟶ $E = hf$ ⟵ frequency (Hz)

Planck's constant

LEARN
U.V. light frees electrons because it has more energy than white light. $E = hf$

It is useful to compare thermionic emission with photoelectric emission.

Thermionic Emission

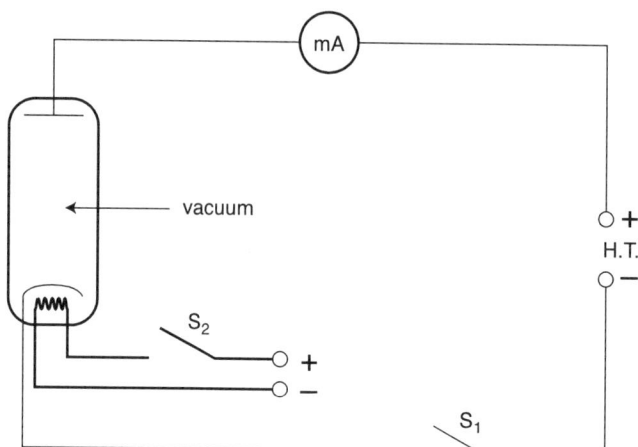

If only S_1 is closed there is no reading on the milliammeter.

If both S_1 and S_2 are closed a current will register on the milliammeter (provided the heater is heating electrons).

When the H.T. supply is reversed and positive charges are heated, there is no current on the milliammeter.

Conclusions

1. Positive charges do not move.

2. Negative charges move. $\boxed{\therefore \text{ CURRENT IS A FLOW OF ELECTRONS}}$

3. Probably electrons are small and light but protons are big and heavy.

4. Nothing happens without the heater so the heater must supply energy to the electrons to allow them to escape from the cathode.

Photoelectric Emission

Instead of using heat as in thermionic emission, U.V. light can be used to supply energy to free electrons from a zinc cathode. They are attracted to the anode and form the photoelectric current.

When the U.V. lamp is switched off, the photoelectric current is zero.

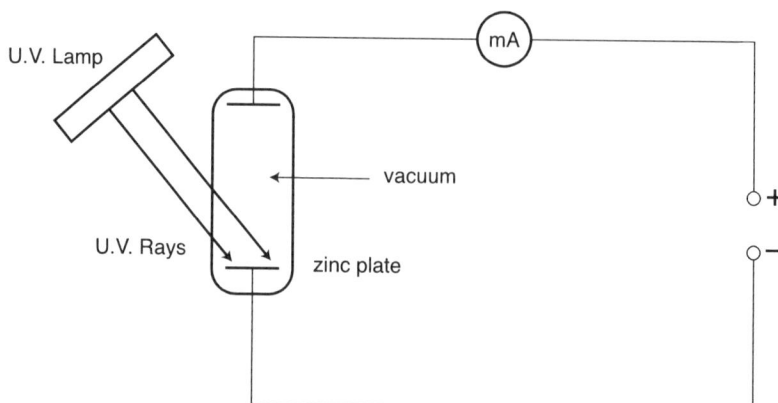

Effect of increasing frequency of U.V. light

As frequency (and so energy) of U.V. light is increased, the photoelectric current increases.

There is a threshold frequency (f_0) below which no current is produced.

photoelectric current

frequency

f_0

Effect of increasing intensity of U.V. light

photoelectric current $\propto$ intensity

Provided that the frequency is greater than the threshold frequency (f_0).

If the frequency is less than the threshold, the intensity can be very large but there will be no current.

photoelectric current

intensity

Discussion and Equations

- The results of the photoelectric effect experiments lead us to think of light as made up of bits of light or **photons**.

- Each individual photon has energy $E = hf$.

- Provided it is above threshold frequency $E = hf_0$, an electron will be freed.

- If the incident photon has energy $E = hf$ which is greater than hf_0, then the electron will escape and have extra energy in the form of kinetic energy.

Energy of incident photon	$hf = hf_0 + \frac{1}{2}mv^2$

minimum energy required to free electron (work function)

kinetic energy of free electron

LEARN
Photoelectric current $\propto$ intensity provided $f > f_0$. $hf = hf_0 + \frac{1}{2}mv^2$

Progress to P.P. 'H' P. Page 122, nos 1–10

Emission   **Absorption**

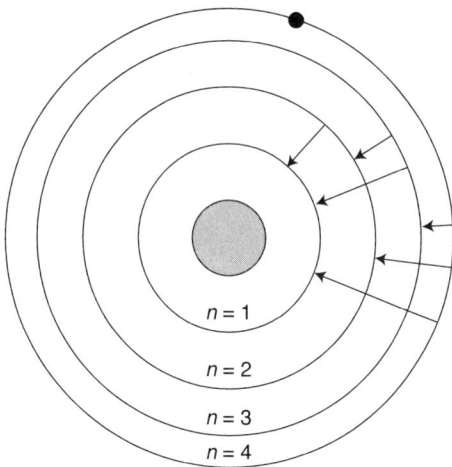

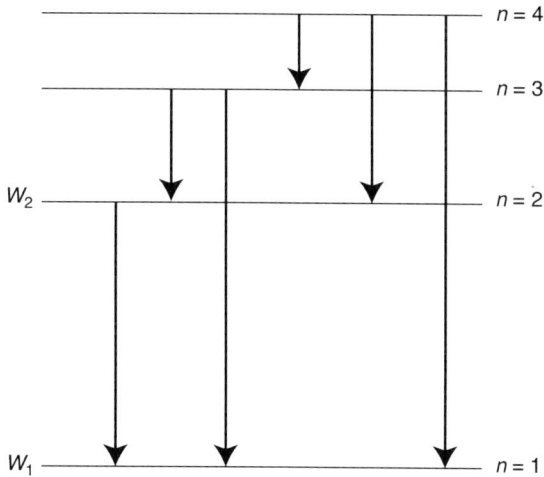

6 spectral lines
e.g., light emitted (2 → 1) with energy

$$hf = W_2 - W_1$$

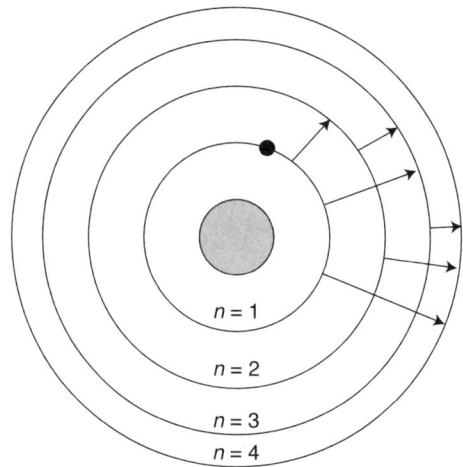

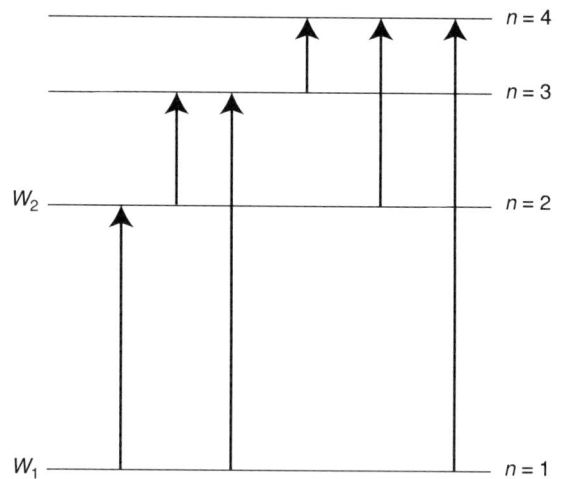

6 spaces (lines) in continuous spectrum
e.g., light absorbed (1 → 2) with energy

$$hf = W_2 - W_1$$

A specific element (like hydrogen)
emits the same frequencies that it **absorbs**.

SPECTRA

The Structure of the Atom is of a central core or nucleus containing protons and neutrons. Electrons orbit this nucleus, like planets round the sun, in discrete energy levels. The electrons can only exist in specific orbits like the 1st or 2nd orbit (or shell) and cannot exist in between.

Bohr postulated this model of electrons in shells, following research by Balmer and others, to explain spectral lines.

Emission Spectra

- When a gas, e.g., hydrogen, in a discharge tube is subjected to high voltage, electrons in the 1st shell absorb energy and jump to a higher energy level, e.g., 4th shell.

- Electrons do not want to be in a higher energy level and try to jump back to a shell closer to the nucleus.

- When this happens they emit energy in the form of light.

The emission diagrams opposite show six possible jumps (called transitions), i.e., six specific frequencies (or wavelengths).

Each element has its own characteristic wavelengths. These can be used to identify the element — sometimes called an optical fingerprint.

Absorption Spectra

When presented with a full range of frequencies, an electron in the 1st shell (opposite) can only absorb certain frequencies corresponding to specific energies to allow it to jump to the 2nd, 3rd, etc., shells. An absorption line in a spectrum is a space where a frequency (or wavelength) is missing. Light from the sun has lines missing (Fraunhofer lines) corresponding to frequencies absorbed by gases in the solar atmosphere.

Ground state: electron in the lowest energy level.
Excited state: electron in a higher energy level (2nd, 3rd, etc., shell).
Ionisation level: electron gains enough energy to escape from the atom.

Turn over for Worked Example on Spectra

LEARN
$W_2 - W_1 = hf$

Progress to P.P. 'H' P. Page 126, nos 1–7

Worked Example on Spectra

Problem

An electron in the 3rd shell of hydrogen jumps to the 1st shell and emits light.

Calculate the frequency of this light.

Energy levels converge to continuum

$n = 4$ ————————— $-1 \cdot 360 \times 10^{-19}$ J

$n = 3$ ————— W_3 ————— $-2 \cdot 416 \times 10^{-19}$ J

$n = 2$ ————————— $-5 \cdot 424 \times 10^{-19}$ J

$n = 1$ ————— W_1 ————— $-21 \cdot 760 \times 10^{-19}$ J

Solution

The problem is solved in two steps.

Step ①
Find the energy of the light.
This is the difference between the energy levels W_3 and W_1.
or
$(21 \cdot 760 - 2 \cdot 416) \times 10^{-19}$ J
$\qquad = 19 \cdot 344 \times 10^{-19}$ J

Step ②
From $E = hf$ then $f = \dfrac{E}{h} = \dfrac{19 \cdot 344 \times 10^{-19}}{6 \cdot 63 \times 10^{-34}} = \underline{\underline{2 \cdot 92 \times 10^{15} \text{ Hz}}}$

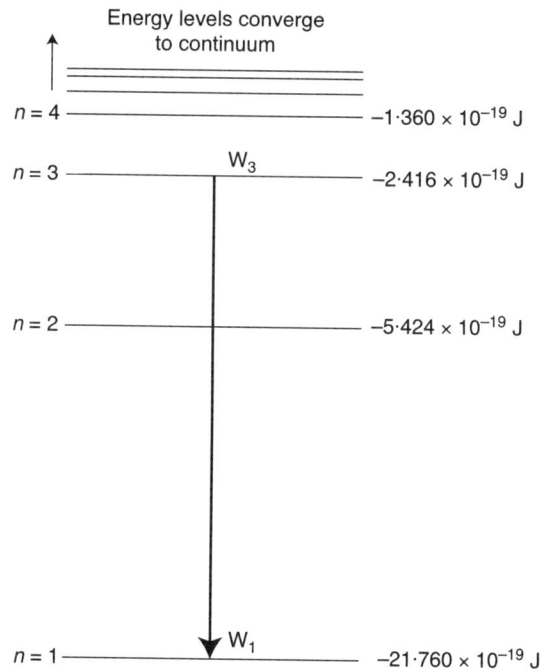

INSIDE A LASER

A helium-neon laser uses two mirrors, M_1 and M_2, as shown.

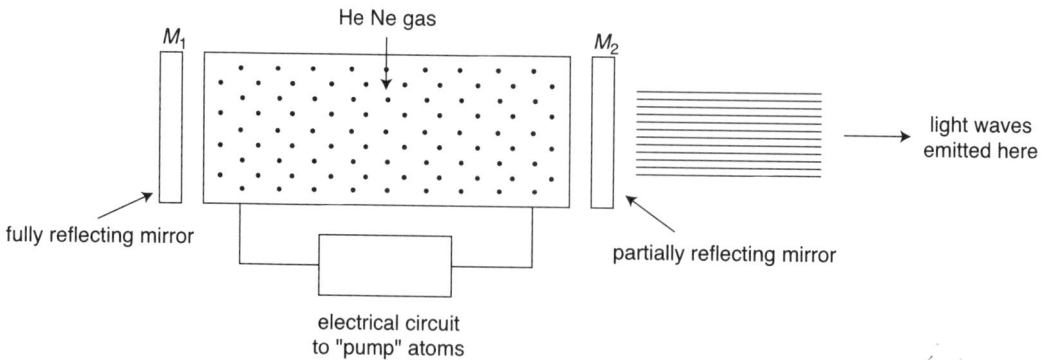

M_1

He Ne gas

M_2

light waves emitted here

fully reflecting mirror

electrical circuit to "pump" atoms

partially reflecting mirror

The mirrors reflect photons of light backwards and forwards between them, stimulating the HeNe gas into further emission of identical photons.

LASERS

Spontaneous emission of light is a random process just like radioactive decay is a random process. Nobody can predict when an electron will jump to a lower energy level.

Stimulated emission of light occurs when an electron transition is triggered by an incident (stimulating) photon, i.e.,

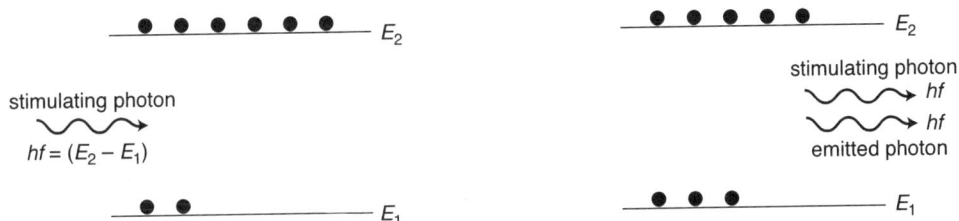

The energy of the stimulating photon (hf) must be the same as the energy gap ($E_2 - E_1$) in order to cause the electron to jump down.

The emitted photon is identical to the original stimulating photon. For each photon in, two come out — hence the name Light Amplification by the Stimulated Emission of Radiation (L.A.S.E.R.).

Two mirrors are used inside a laser (see diagram opposite).

The emitted laser light is:

MONOCHROMATIC — all photons have the same frequency.
COHERENT — all photons are in phase.
INTENSE — all photons are in phase and concentrated into a small area.
PARALLEL — due to the plane mirrors (only light striking the mirrors at 90° is reflected and amplified).

Typical intensity

For a laser of power 0·1 mW forming a beam of radius 0·75 mm:

$$\text{Intensity} = \frac{P}{A} = \frac{0.1 \times 10^{-3}}{\pi r^2} = \frac{0.1 \times 10^{-3}}{3.142(0.75 \times 10^{-3})^2} = \underline{\underline{56.6 \ Wm^{-2}}}$$

This level of intensity can cause eye damage.

Progress to P.P. 'H' P. Page 128, nos 1–7

SEMICONDUCTORS

Semiconductor Theory

Materials can be divided into three broad categories according to their electrical properties.

Conductors have many free electrons which can move easily, e.g., silver, copper, graphite.

Insulators have very few free electrons which cannot move easily, e.g., wood, paper, rubber.

Between conductors and insulators is a group of materials called **semiconductors**. The resistance of a semiconductor can easily change according to the number of electrons that are free to move. When a pure (intrinsic) semiconductor, such as germanium, is heated, electrons can be freed to conduct an electric current.

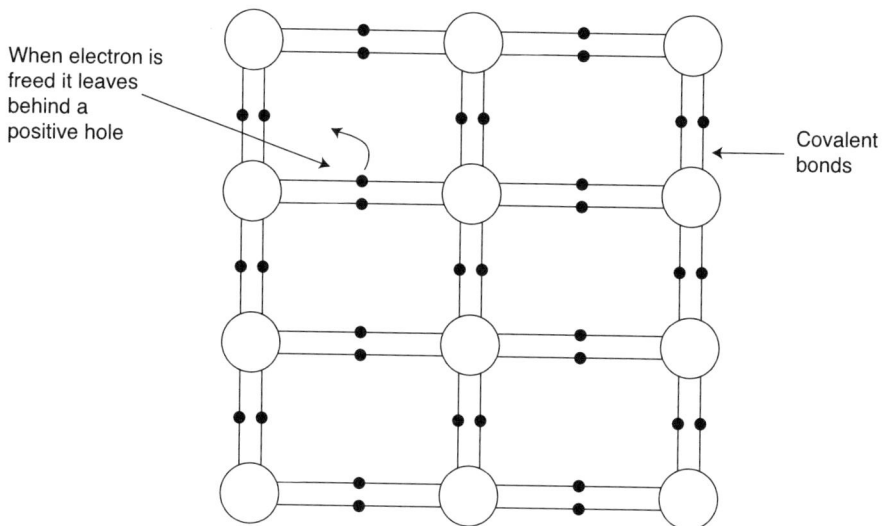

When an electron is freed to conduct, it leaves behind a hole which is positively charged. Thus electron-hole pairs are produced. Conduction can now take place by electrons moving towards the positive potential and holes moving towards the negative potential. A hole moves when the atom captures an electron from the covalent bond of a neighbouring atom.

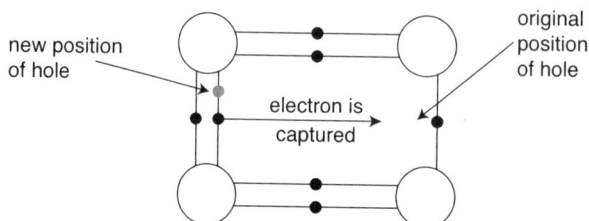

The warmer the semiconductor the more electron-hole pairs are produced and the greater the current. Electron-hole pairs can be produced by light energy (the photo-electric effect). The resistance of the semiconductor decreases as the number of electron-hole pairs increases.

Doped Semiconductors

As well as the action of heat and light, the resistance of a semiconductor can be reduced by "doping" the intrinsic semiconductor with a minute trace of impurity.

n-type Semiconductor

When a group V (non-metal) impurity such as arsenic with five electrons in the outer shell is added to the germanium crystal lattice, it very nearly fits. Four electrons are bonded in the covalent bonds and one electron is left over free to move through the germanium.

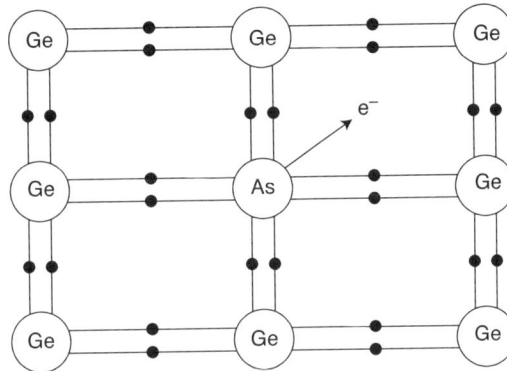

The group V impurities, e.g., arsenic, antimony, phosphorus, are called donor impurities. Room temperature produces electron-hole pairs in addition to the donated electrons. In *n*-type semiconductor there is an excess of electrons (holes are in the minority). In this case, the electrons are called majority carriers and holes are called minority carriers.

p-type Semiconductor

When a group III (metal) impurity such as indium with three electrons in the outer shell is added to the germanium crystal lattice, it very nearly fits. All three electrons are bonded in covalent bonds but there is a space where an electron should be — a positive hole.

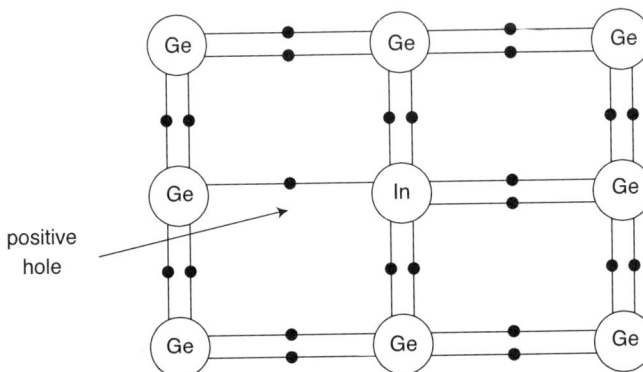

The group III impurities, e.g., boron, indium, gallium, are called acceptor impurities. Room temperature produces electron-hole pairs in addition to the positive holes created by the acceptor atoms. In *p*-type semiconductor there is an excess of positive holes (electrons are in the minority). In this case, the electrons are the minority carriers while the holes are the majority carriers.

The *p-n* junction diode

A *p–n* junction is not a piece of *p*-type semiconductor stuck to a piece of *n*-type semiconductor. It is a single crystal of germanium specially grown so that one side is doped *p*-type and the other side is doped *n*-type.

When the *p–n* junction is formed, some electrons migrate across the junction and fill holes in the *p*-type germanium. At the same time, some positive holes migrate across the junction and combine with electrons in the *n*-type germanium.

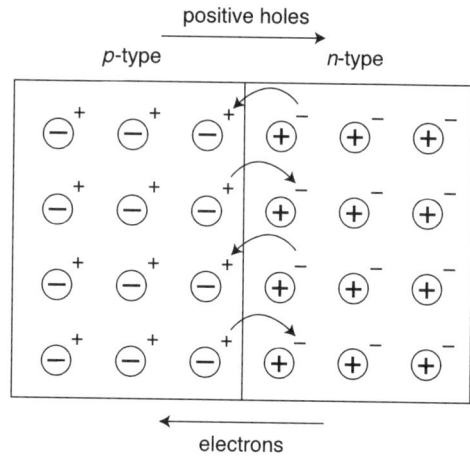

⊕ and ⊖ indicate fixed charge centres

+ and − indicate mobile charge carriers:
− indicate electrons
+ indicate positive holes

This leaves a narrow band of negatively charged atoms on the *p*-type side of the junction and a narrow band of positively charged atoms on the *n*-type side of the junction. The negative fixed charge centres repel any further migration of electrons while the positive fixed charge centres repel any further migration of positive holes. This layer, called a depletion layer, forms a potential barrier which prevents any flow of charge across the junction

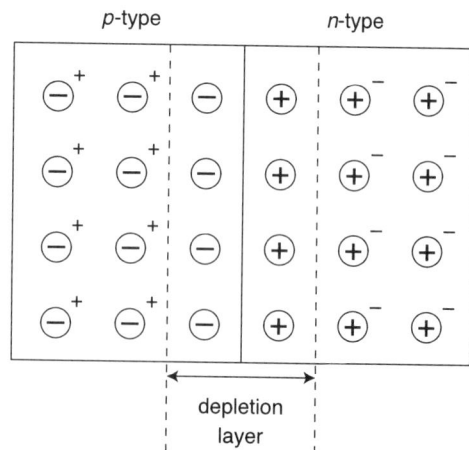

Reverse bias

When a battery is connected across the *p–n* junction as shown, the potential barrier increases and no current flows in the circuit.

The *p–n* junction is said to be back biased or reverse biased.

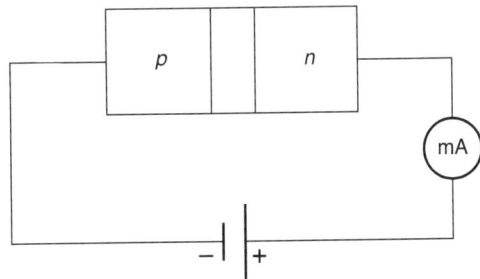

Forward bias

When a battery is connected across the *p–n* junction as shown, the depletion layer disappears and current flows easily round the circuit. Electrons flow from *n* to *p* and holes flow from *p* to *n*.

The *p–n* junction is said to be forward biased.

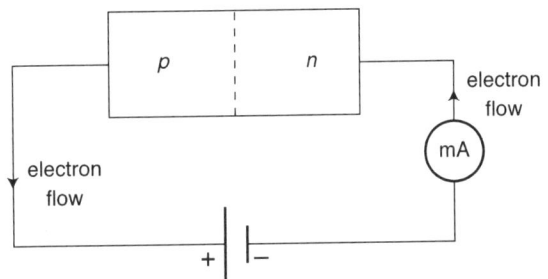

The Light Emitting Diode

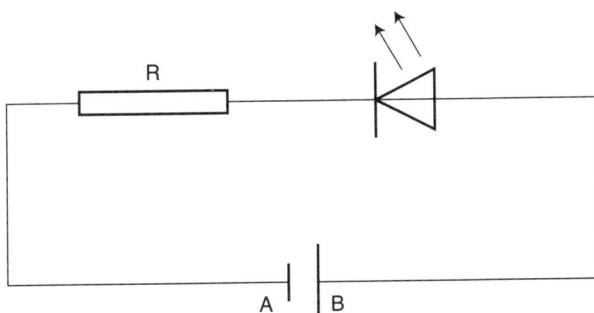

When the *p–n* junction is forward biased, electrons flow from A to B. Some of the electrons recombine with holes at the junction emitting energy. In most semiconductors this is in the form of heat but in certain semiconductors the energy is emitted as photons of light.

One electron filling one hole (recombination) emits one photon of light of energy $E = hf$ where f is the frequency of the light emitted.

Progress to P.P. 'H' P. Page 129, nos 1–5

The Photodiode

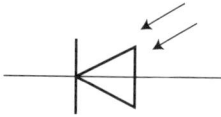

The photodiode is the opposite of the L.E.D. (and so is the symbol).

When light strikes the $p{-}n$ junction, each photon of light frees an electron, leaving a hole behind.

It can be used in two modes:

Photovoltaic mode

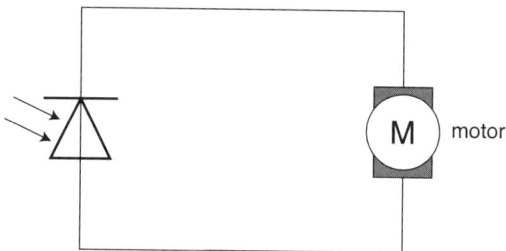

In photovoltaic mode, the photodiode may be used to supply power to a load since it behaves like a small battery (solar cell).

Photons of light free electrons, creating electron-hole pairs and causing a voltage. The more photons of light, the more electron-hole pairs and the bigger the voltage.

Photoconductive mode

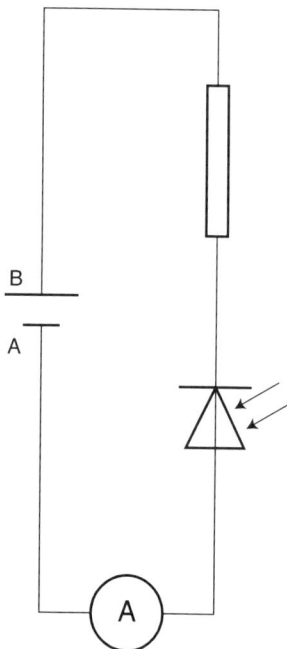

In photoconductive mode, the photodiode is connected in a circuit **reverse biased** so that no current flows.

When light shines on the photodiode, electrons are freed, creating electron-hole pairs and a current now flows in the circuit (from A to B).

This (leakage) current is proportional to the intensity of the incident light. Therefore the resistance of the photodiode drops as light intensity increases.

The switching action of a reverse-biased photodiode is extremely fast.

Progress to P.P. 'H' P. Page 131, nos 6–8

METAL OXIDE SEMICONDUCTOR FIELD EFFECT TRANSISTORS (MOSFETS)

Although there are two types of MOSFETs in *n*-channel and *p*-channel, only the *n*-channel enhancement MOSFET will be dealt with in this course.

Structure of the MOSFET

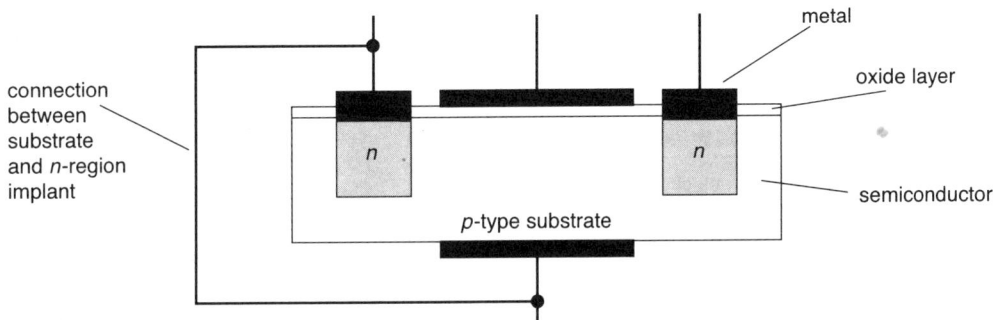

The structure of the MOSFET (above and below) shows a slab of *p*-type semiconductor called the **substrate** with two *n*-type **implants** at each end. A thin **oxide layer** (silicon dioxide) covers the substrate and both implants. Three metal contacts are made — one to each implant and one to the oxide layer. These three contacts are called the **gate**, the **source** and the **drain**.

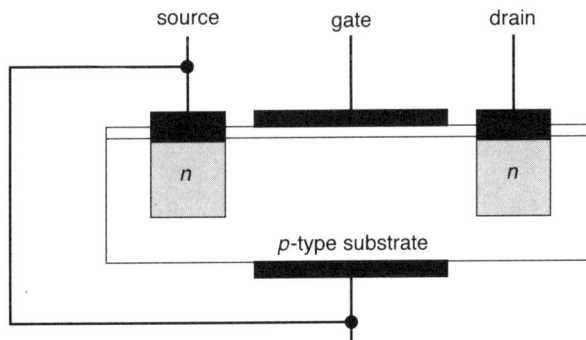

In any circuit, the idea is to send a current from the source (source of electrons) to the drain (drain of electrons). Current will only flow provided the gate permits it.

Switching the MOSFET on

When a supply is connected between the source and the drain (gate not connected), no current will flow.

When a second battery (in addition to the supply) is now connected between the source and the gate (gate positive with respect to the source), a current will flow from source to drain (provided gate voltage is high enough — usually about 2 V).

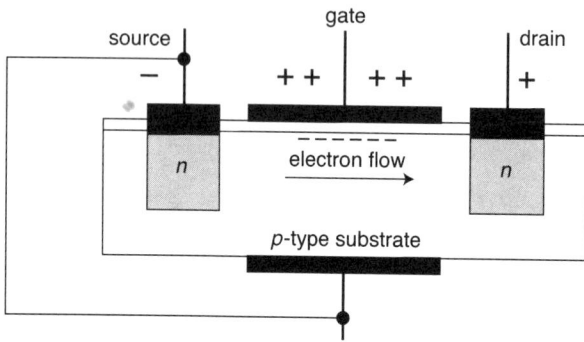

The positive charge on the gate attracts electrons from the p-type substrate.

[N.B. Heat forms electron-hole pairs in the p-type substrate.]

This channel of electrons is called an n-channel.

MOSFET circuits

In MOSFET circuits, the diagram above would be replaced with a circuit symbol. Turning the diagram through 90° shows where the circuit contacts are for gate, source and drain.

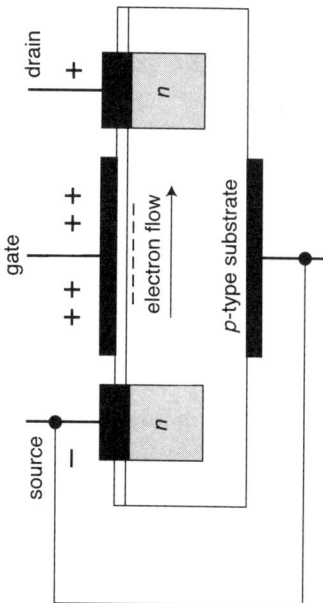

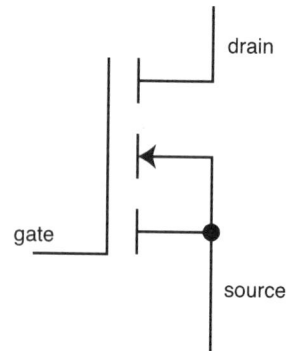

The gate voltage must be above the threshold (e.g., 2 V) to switch the MOSFET on.

The MOSFET as a switch

Provided the gate voltage (V_{GS}) is greater than the threshold voltage (about 2 V) the MOSFET is switched on.

A current now flows from source (S) to drain (D). The drain current, I_D, passes through the load resistor. The load resistor could be any device: a lamp, buzzer, motor, etc.

There are two ways to increase the drain current:

1. Increase V_{GS} (beyond the threshold).

2. Increase V_{DS}.

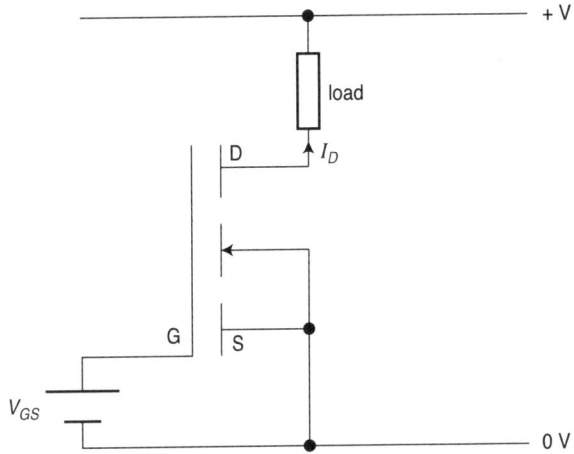

The MOSFET as an amplifier

The MOSFET can also be used as an amplifier.

R_1 and R_2 form a voltage divider to ensure V_1 is above the threshold voltage.

As V_1 increases the bulb gets brighter.

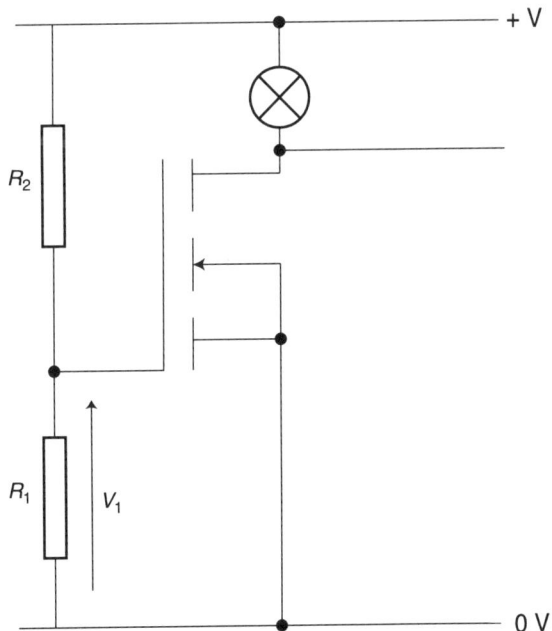

LEARN
The gate voltage must be above the threshold (e.g., 2 V) to switch the MOSFET on.

Progress to P.P. 'H' P. Page 131, nos 9–12

Thomson Model (plum pudding)

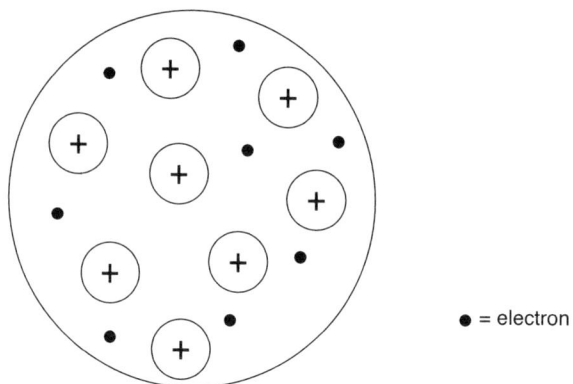

● = electron

Geiger-Marsden Experiment

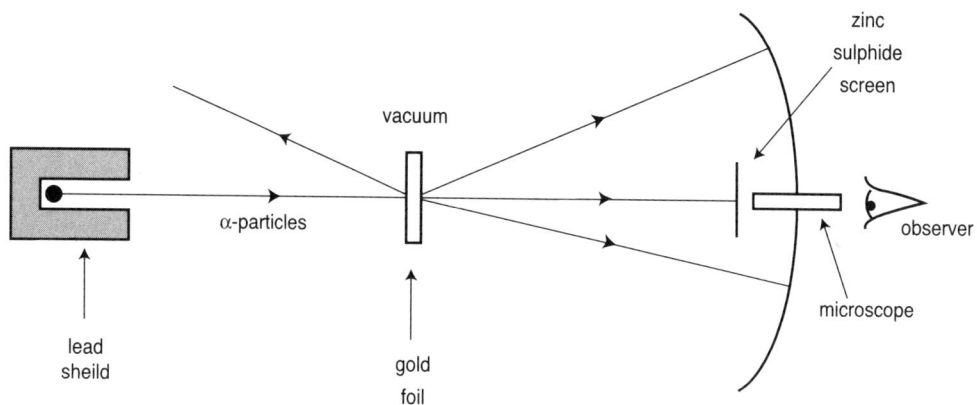

zinc
sulphide
screen

vacuum

α-particles

observer

lead
sheild

gold
foil

microscope

The telescope and zinc sulphide screen are mounted on a moveable turntable which can be rotated in an arc around the gold foil.

CHAPTER 14

NUCLEAR REACTIONS

MODELS OF THE ATOM

As scientists did more and more research into the atom, new particles were discovered and the picture of the atom began to change.

The Plum Pudding Model

At the end of the 19th century, Thomson discovered the electron and, coupled with the results of other experiments, he put forward the plum pudding model. He knew at the time that there were two kinds of charge and he knew that electrons could move (cathode rays), but the positive part of the atom did not move (no positive rays in a glass tube). His model described the atom as a positive material shaped like a sphere with electrons distributed throughout this positive sphere to neutralise it. This model gets its name from the obvious resemblance to currants in a plum pudding. Thomson could explain charging with this model and also give a reasonable explanation of why there were cathode rays but no positive rays. However, it wasn't long before other researchers were beginning to question this model and put forward their own ideas.

The Rutherford Model

Rutherford was one of the first men to question Thomson's plum pudding model. He asked his colleague Geiger to work with Marsden and try bombarding a thin gold leaf with a beam of alpha particles. Rutherford wanted to see if the alpha particles would be deflected.

The experimental set-up is shown in the diagram opposite. A collimated beam of alpha particles from a radium source is fired at a thin film of gold.

Some of the beam passes straight through the gold and strikes a zinc sulphide screen, producing flashes of light. The number of flashes, and hence the number of alpha particles, can be counted by observing the screen through a microscope.

Geiger and Marsden found that most of the beam travelled straight through as expected but some of the alpha particles were deflected through various angles and a few were actually deflected through large angles, i.e., back the way they had come. So astounded by these results was Rutherford that he said, "It was as though you had fired a 15-inch shell at a sheet of paper and it had come back and hit you!".

Rutherford Model

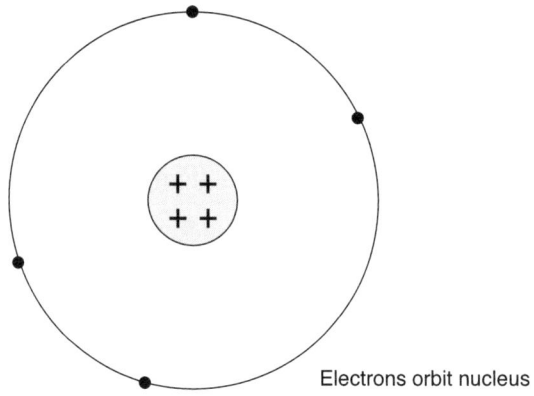

Electrons orbit nucleus

α-paraticle scattering

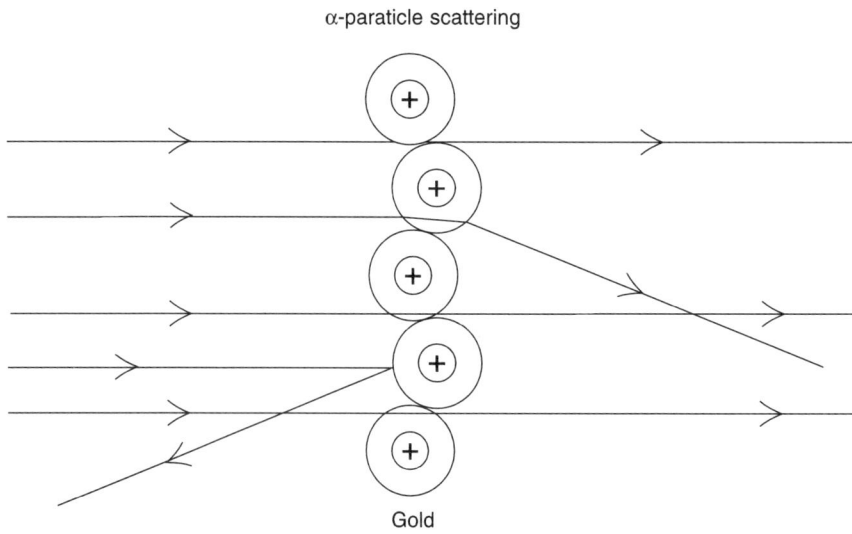

Gold

Bohr Model

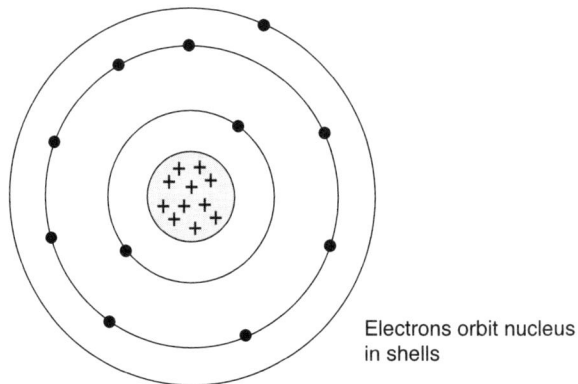

Electrons orbit nucleus
in shells

From these results, Rutherford suggested that although the atom occupied a certain volume, most of the volume was space and all the mass of the atom was concentrated in a small centre core or nucleus which was positively charged.

Spinning around this nucleus at the extremity of the atom were the electrons.

When an alpha particle came very close to a nucleus, the repulsion between the positively charged alpha particle and the positively charged nucleus caused the alpha particle to be deflected. The closer the alpha particle is to the nucleus, the bigger the deflection of the beam. An alpha particle is deflected through a large angle when it makes a head-on collision with a nucleus.

Thus the Rutherford atom seemed to explain everything very well.

Even today, although we know a lot more about the atom than Rutherford, we still talk about a nucleus with electrons outside it.

The Bohr Model

One major argument against Rutherford's atom was that electrons spinning around the nucleus should radiate energy. Having less energy, they would move closer to the nucleus and continue to radiate energy until they actually spiralled into the nucleus itself. Neils Bohr put forward the postulate that electrons could only spin around the nucleus in certain orbits or shells as he called them. In these shells, the electrons could safely orbit the nucleus without radiating energy. The work of Balmer and other researchers helped to back up Bohr's idea and the Bohr model became generally accepted. When Bohr put forward his model (1912), it was based on two particles and it wasn't until 1932, when Chadwick discovered the neutron, that the model was seriously questioned. In the same year, Anderson, working in America, discovered a particle that made the same kind of track as a beta particle but it bent the "wrong way" in a magnetic field. This meant that he had discovered a particle the same size as an electron but positively charged. The particle was the anti-electron or **positron** predicted by Dirac. To date, there have been many more models which try to explain new facts or even new particles — the mathematical model and the liquid-drop model to name only two.

Mass and Atomic Numbers

The atom opposite is sodium, symbol $^{23}_{11}$Na.

$$\text{Mass number} \rightarrow 23 \longrightarrow \text{23 particles in the nucleus.}$$
$$\text{Na}$$
$$\text{Atomic number} \rightarrow 11 \longrightarrow \text{11 protons in the nucleus.}$$

Subtracting 11 from 23 gives 12 = the number of neutrons in the nucleus.

Progress to P.P. 'H' P. Page 133, nos 1–3

NUCLEAR REACTIONS

Alpha Emission

When an alpha particle is emitted from a nucleus, the mass of the nucleus is reduced by 4 a.m.u. and the charge is reduced by 2 (positive charges).

e.g., $\qquad$ $^{232}_{90}$ Th $\longrightarrow$ $^{228}_{88}$ Ra $\qquad$ + $\qquad$ $^{4}_{2}$ He

Notice that the sum of the mass numbers on the right of the equation is the same as the mass number on the left of the equation and the sum of the atomic numbers on the right of the equation is the same as the atomic number on the left of the equation.

$\qquad$ I.e., $\qquad$ 232 $\qquad$ = $\qquad$ 228 $\quad$ + $\qquad$ 4

$\qquad\qquad\qquad\quad$ 90 $\qquad$ = $\qquad$ 88 $\quad$ + $\qquad$ 2

Beta Emission

The same rule for alpha emission also applies for beta emission. An example of beta emission is:

$\qquad$ $^{228}_{88}$ Ra $\longrightarrow$ $^{228}_{89}$ Ac $\qquad$ + $\qquad$ $^{0}_{-1}$ e

In beta emission, a neutron in the nucleus breaks down into a proton which stays in the nucleus and an electron which is emitted as a beta particle. Therefore, the number of particles in the nucleus does not change, which explains no change in mass number — but the number of protons increases by one, i.e., from 88 to 89.

Gamma Emission

After emitting an alpha or beta particle, sometimes the nucleus is left in an excited state. It can lose this excess energy by emitting a gamma ray.

Mass and Energy

When Anderson discovered the positron, it was by observing how a gamma-ray photon reacted when it struck a lead target.

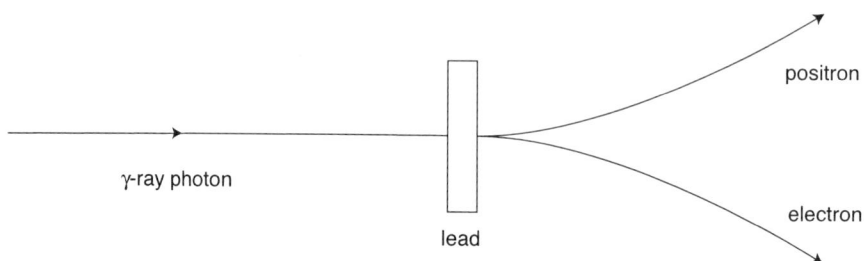

As well as discovering a new particle, the experiment shows pure energy being converted into mass. The opposite has since been observed: when an electron meets a positron, they annihilate each other and form pure energy. The total sum of mass and energy before the collision is equal to that after the collision.

Properties of α, β and γ radiation

Radiation	Description	Mass (a.m.u.)	Charge	Ionising power [ion pairs per cm of track]	Speed	Stopped by
α	Helium nucleus $^{4}_{2}\text{He}$	4	+2	2×10^5	comparatively slow ~ 10% the speed of light	a sheet of paper
β	very fast electron $^{0}_{-1}e$	$\dfrac{1}{1836}$	−1	~100	~ 90% the speed of light	a few millimetres of aluminium
γ	high frequency high energy electro-magnetic radiation $^{0}_{0}\gamma$	0	0	very few (produces ions by secondary processes)	the speed of light = C = 3×10^8 m s^{-1}	several centimetres of lead

Progress to P.P. 'H' P. Page 134, nos 4–10

FISSION AND FUSION

Fission

Fission occurs when a nucleus splits to form two smaller nuclei (and normally other small particles). Although fission can be spontaneous, it is also possible to induce fission with a neutron.

Typical (induced) fission

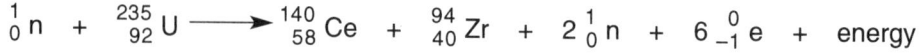

$$^{1}_{0}n \; + \; ^{235}_{92}U \longrightarrow \; ^{140}_{58}Ce \; + \; ^{94}_{40}Zr \; + \; 2\,^{1}_{0}n \; + \; 6\,^{0}_{-1}e \; + \; energy$$

Energy is formed because mass is lost.

Total mass before reaction

$$
\begin{aligned}
\text{mass of U 235} &= 235 \cdot 044 \\
\text{mass of n} &= \underline{\;\;1 \cdot 009\;\;} \\
&\quad\; 236 \cdot 053 \text{ a.m.u.}
\end{aligned}
$$

Total mass after reaction (ignoring the electrons).

$$
\begin{aligned}
\text{mass of Ce} &= 139 \cdot 905 \\
\text{mass of Zr} &= \;\;93 \cdot 906 \\
\text{mass of 2 n} &= \underline{\;\;\;2 \cdot 018\;\;} \\
&\quad\; 235 \cdot 829 \text{ a.m.u.}
\end{aligned}
$$

Mass Defect

$$
\begin{aligned}
\text{mass defect} &= 236 \cdot 053 - 235 \cdot 829 \\
&= \underline{\;\;0 \cdot 224 \text{ a.m.u}\;\;}
\end{aligned}
$$

This can be converted into kg using 1 a.m.u. $= 1 \cdot 660 \times 10^{-27}$ kg.

$$\therefore \quad \text{mass defect (kg)} = 0 \cdot 224 \times 1 \cdot 660 \times 10^{-27}$$
$$= 3 \cdot 7184 \times 10^{-28} \text{ kg}$$

Energy Produced

This lost mass (mass defect) is converted into energy.

$$E = m\,c^2$$
$$E = 3 \cdot 7184 \times 10^{-28} \,(3 \times 10^{8})^2$$
$$E = \underline{\underline{3 \cdot 35 \times 10^{-11} \text{J}}}$$

In a chain reaction, one (or both) of the neutrons produced in this reaction goes on to create a further fission reaction.

Fusion

Fusion occurs when two small nuclei fuse together to form a larger nucleus.

Typical fusion

$$^2_1H \; + \; ^2_1H \longrightarrow \; ^3_2He \; + \; ^1_0n \; + \; energy$$

Energy is formed because mass is lost.

Total mass before reaction

mass of ^2_1H $\quad = 3 \cdot 342 \times 10^{-27}$

mass of ^2_1H $\quad = \underline{3 \cdot 342 \times 10^{-27}}$

$\qquad\qquad\qquad\qquad 6 \cdot 684 \times 10^{-27}$ kg

Total mass after reaction

mass of He $\quad = 5 \cdot 004 \times 10^{-27}$

mass of n $\quad = \underline{1 \cdot 674 \times 10^{-27}}$

$\qquad\qquad\qquad 6 \cdot 678 \times 10^{-27}$ kg

Mass Defect

mass defect $\quad = (6 \cdot 684 - 6 \cdot 678) \times 10^{-27}$

$\qquad\qquad\qquad = \underline{\underline{6 \cdot 0 \times 10^{-30} \text{ kg}}}$

Energy Produced

This lost mass (mass defect) is converted into energy.

$$E = m c^2$$
$$E = 6 \cdot 0 \times 10^{-30} (3 \times 10^8)^2$$
$$E = \underline{\underline{5 \cdot 4 \times 10^{-13} \text{ J}}}$$

- Fusion reactions like this produce energy on the sun and other stars.
- Energy produced in fission and fusion reactions results in greater kinetic energy of the products.

Progress to P.P. 'H' P. Page 135, nos 1–12

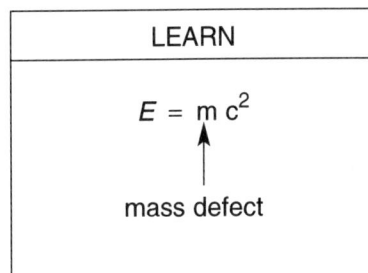

LEARN
$E = m c^2$
↑
mass defect

CHAPTER 15

DOSIMETRY AND SAFETY

DOSIMETRY EQUATIONS

Activity

The activity of a source is measured in becquerels where one becquerel (Bq) is defined as one decay per second.

activity / Bq ⟶ $A = \dfrac{N}{t}$ ⟵ no. of nuclei decayed
⟵ time / s

e.g., 100 Bq = 100 decays per second.

Absorbed Dose

The absorbed dose is defined as the energy absorbed per unit mass of the absorbing material.

absorbed dose / gray ⟶ $D = \dfrac{E}{m}$ ⟵ energy / J
⟵ mass / kg

The unit of absorbed dose is the gray (gy) where 1 gray = 1 joule per kilogram.

Quality factor

Is a certain dose of α-particles more or less harmful than the same dose of γ-rays?

To answer this, scientists must look at what the radiation does to the human body.

The γ-rays pass through the body but the α-particles do not.

The α-particles do more damage than the same dose of γ-rays — in fact, twenty times more damage — since they create much more ionisation than γ-rays.

This leads to a scale of damage which is a reflection of the biological effect the particular radiation has on the human body.

The scale is simply a number (no units) called the Q-value for the radiation. As can be seen from the table, a dose of protons is 10 times more harmful than the same dose of β-particles.

The risk of biological harm from an exposure to radiation depends on:

(a)   the absorbed dose;

(b)   the kind of radiation, e.g., α, β, etc.;

(c)   the body organs or tissues exposed.

Type and energy of radiation	Q
X and γ-rays	1
β-particles	1
α-particles	20
protons	10
neutrons: slow	2·3
neutrons: fast	10

LEARN
$A = \dfrac{N}{t}$
$D = \dfrac{E}{m}$

Progress to P.P. 'H' P. Page 138, nos 1–8

Dose Equivalent

The dose equivalent takes the Q value into account.

The dose equivalent is the product of the dose and the Q value.

dose equivalent / sieverts (Sv) $\longrightarrow$ $\boxed{H = DQ}$ — dose / gy

— Q value

Worked Example

Problem

What is the dose equivalent of a dose of 5 m gy of

(a) γ-rays,

(b) α-particles?

Solution

(a) γ-rays
$$H = DQ$$
$$H = 5 \times 10^{-3} \times 1$$
$$\underline{\underline{H = 5 \text{ mSv}}}$$

(b) α-particles
$$H = DQ$$
$$H = 5 \times 10^{-3} \times 20$$
$$\underline{\underline{H = 100 \text{ mSv}}}$$

Dose Equivalent Rate

The dose equivalent rate is the dose equivalent per unit time.

dose equivalent rate / Sv per hour
or Sv per year
etc.
$\longrightarrow$ $\boxed{\dot{H} = \dfrac{H}{t}}$ — dose equivalent / Sv

— time / hour or year, etc.

The effective dose equivalent (Sv) takes account of the different susceptibilities to harm of the tissues being irradiated and is used to indicate the risk to health from exposure to ionising radiation.

Background Radiation

The average annual effective dose equivalent which a person in the U.K. receives due to natural sources is approximately 2 mSv.

Source of background radiation	Annual dose equivalent
Cosmic radiation	0·3 mSv
Radioactivity from rocks, soils and building materials	0·3 mSv
Radioactivity in the human body	0·4 mSv
Inhaled radon and its daughter products	1·0 mSv
Total	2·0 mSv

However, this is only an average. The dose equivalent from a chest X-ray is 2·0 mSv.

Exposure Limits

Annual effective dose equivalent limits have been set for exposure to radiation for the general public and higher limits for workers in certain occupations.

> For **members of the public**, the dose equivalent should not exceed **5 mSv** in a year (in addition to background) and should not exceed **1 mSv** in a year as a long-term average.
>
> For **workers exposed as part of their employment**, the dose equivalent should not exceed **50 mSv** in a year (in addition to background).

The dose equivalent rate can be reduced by

(a)   shielding or

(b)   increasing the distance from the source.

Progress to P.P. 'H' P. Page 139, nos 9–20

Progress to P.P. 'H' P. Page 140, nos 1–3

LEARN
$H = DQ$
$\dot{H} = \dfrac{H}{t}$

Apparatus

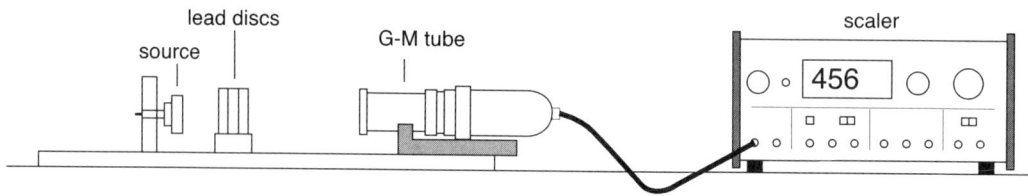

Graph

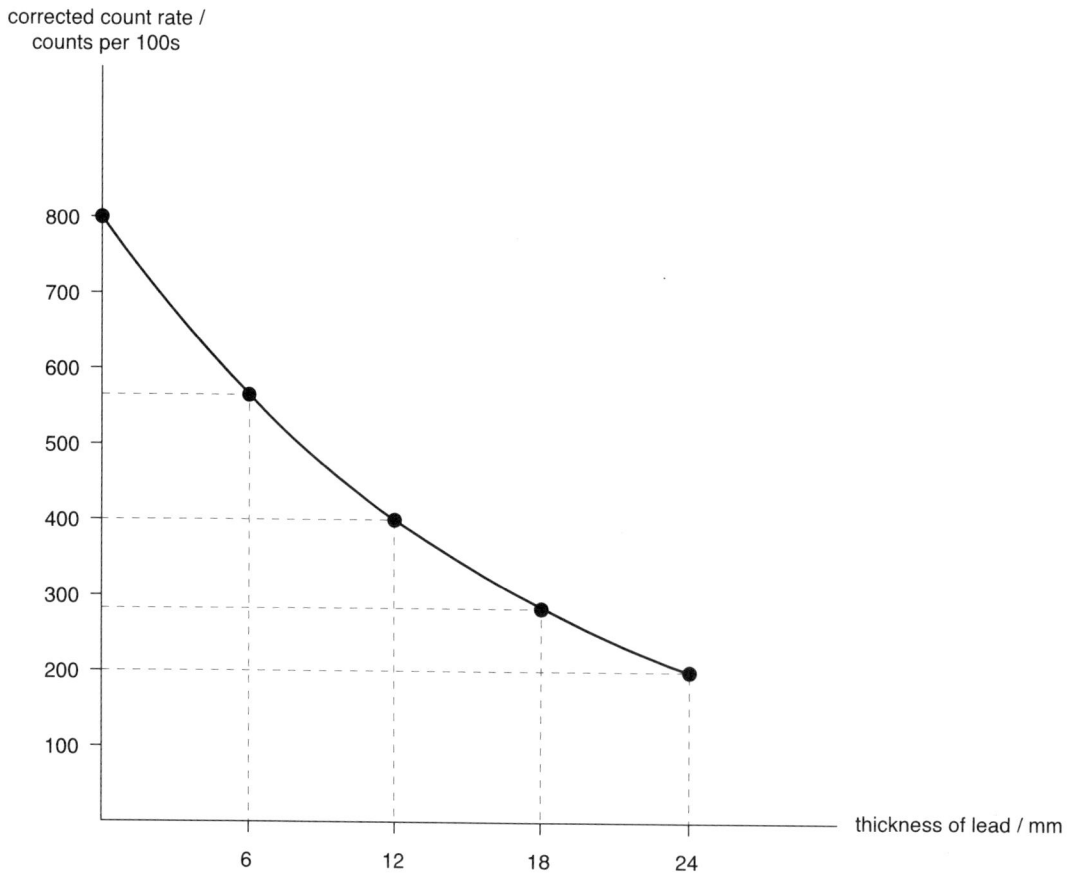

HALF-VALUE THICKNESS OF LEAD FOR GAMMA RAYS

Object

To find the thickness of lead required to reduce the intensity of the beam to half its original value.

Method

The apparatus shown opposite was set up with no lead.

Lead discs were inserted in the space between the source and the G-M tube. The scalar counter was set to count automatically for 100 seconds and then switch off. Three counts were taken for each thickness of lead and the average calculated. Background count rate (per 100s) was measured and subtracted from the average count rate — see results below.

Results

Background count per 100s = 33, 36, 37
Mean background count per 100s = 35

Thickness of lead / mm	Counts per 100s			Mean Counts per 100s	Corrected counts / 100s
0	835	841	829	835	800
6	600	610	590	600	565
12	430	440	436	435	400
18	320	315	318	318	283
24	245	234	226	235	200

Conclusion

A graph of corrected count rate against thickness of lead is plotted opposite. From the graph:

the half-value thickness of lead for gamma rays = 12 mm.

Progress to P.P. 'H' P. Page 141, nos 1–3

UNCERTAINTIES

The measurement of any physical quantity is liable to uncertainty (error).

RANDOM UNCERTAINTY

Four students are working on an experiment which involves reading a voltmeter. When it is time to read the voltmeter a different student takes the reading.

Kieran takes the first reading and reads	2·0 V	Sharon takes the third reading and reads	6·0 V
Rafié takes the second reading and reads	4·1 V	Roanna takes the fourth reading and reads	8·1 V

The teacher tells them that it is better if one student takes all the readings on the voltmeter so that the same person is looking at the needle at the same angle every time.

When the teacher allows a new student, Mari-Elena, to take all four readings she gets: 2·0 V, 4·0 V, 6·0 V, 8·0 V.

Reading Error

Analogue Instrument

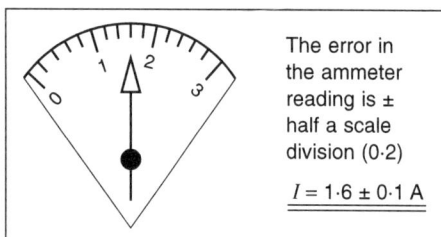

The error in the ammeter reading is ± half a scale division (0·2)

$$I = 1.6 \pm 0.1 \text{ A}$$

Digital Instrument

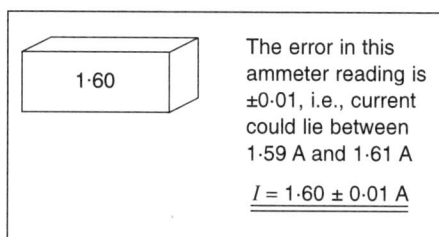

1·60

The error in this ammeter reading is ±0·01, i.e., current could lie between 1·59 A and 1·61 A

$$I = 1.60 \pm 0.01 \text{ A}$$

Percentage Error

The % error in the analogue meter is
$$\frac{0.1}{1.6} \times 100\% = 6.25\%$$

The % error in the digital meter is
$$\frac{0.01}{1.60} \times 100\% = 0.63\%$$

The Mean

The mean is the best estimate of a "true" value of the quantity being measured. Jennifer and Rachael are taking readings of background count rate by measuring the counts every 100 seconds. After the first five readings, Jennifer gets fed up and goes off with these five results. Rachael is more patient and continues for another five readings.

Jennifer's results
35, 31, 33, 36, 33
$$\text{mean} = \frac{35 + 31 + 33 + 36 + 33}{5}$$
$$= 33.6 \text{ counts per 100 s}$$

Rachael's results
35, 31, 33, 36, 33, 35, 36, 35, 34, 33
$$\text{mean} = \frac{35 + 31 + 33 + 36 + 33 + 35 + 36 + 35 + 34 + 33}{10}$$
$$= 34.1 \text{ counts per 100 s}$$

The approximate random error in the mean

Jennifer's approx. random error
$$= \frac{\text{max value} - \text{min value}}{\text{no. of readings}}$$
$$= \frac{36 - 31}{5} = 1.0$$
Jennifer's answer = 33·6 ± 1·0 counts per 100 s

Rachael's approx. random error
$$= \frac{\text{max value} - \text{min value}}{\text{no. of readings}}$$
$$= \frac{36 - 31}{10} = 0.5$$
Rachael's answer = 34·1 ± 0·5 counts per 100 s

Increasing the number of results cuts down the error in the mean so Rachael's answer is more accurate.

Combining Errors

Where an experiment involves the measurement of more than one physical quantity the % error in the answer is approximately the same as the largest % error in the measurements, e.g., in Ohm's Law, $R = \frac{V}{I}$

If reading error in V is 10% and reading error in I is 5% then error in R is 10%

> **LEARN**
>
> Approximate random error
> $$= \frac{\text{max value} - \text{min value}}{\text{number of readings}}$$
> % error
> $$= \frac{\text{absolute error}}{\text{actual measurement}} \times 100\%$$

Progress to P.P. 'H' P. Page 142, nos 1–6

Printed by Bell & Bain Ltd., Glasgow, Scotland, U.K.